AMERICAN SHEEP

American Sheep

A Cultural History

BRETT BANNOR

The University of Georgia Press
ATHENS

This publication is made possible in part through a grant from the Bradley Hale Fund for Southern Studies

© 2024 by the University of Georgia Press
Athens, Georgia 30602
www.ugapress.org
All rights reserved
Designed by Erin Kirk
Set in Adobe Caslon Pro
Printed and bound by Sheridan Books
The paper in this book meets the guidelines for permanence and durability of the Committee on Production Guidelines for Book Longevity of the Council on Library Resources.

Most University of Georgia Press titles are available from popular e-book vendors.

Printed in the United States of America
24 25 26 27 28 C 5 4 3 2 1

Library of Congress Cataloging-in-Publication Data

Names: Bannor, Brett, author.
Title: American sheep : a cultural history / Brett Bannor.
Description: Athens : The University of Georgia Press, [2024] | Includes bibliographical references and index.
Identifiers: LCCN 2024012266 (print) | LCCN 2024012267 (ebook) | ISBN 9780820367163 (hardback) | ISBN 9780820367170 (ebook) | ISBN 9780820367187 (epub)
Subjects: LCSH: Sheep—United States—History.
Classification: LCC SF375.4.A1 B36 2024 (PRINT) | LCC SF375.4.A1 (EBOOK) | DDC 636.300973—dc23/eng/20240603
LC record available at https://lccn.loc.gov/2024012266
LC ebook record available at https://lccn.loc.gov/2024012267

Contents

Acknowledgments vii

Preface xi

Introduction. "The Protection of His Naked Body against the Inclement Elements" 1

Chapter 1. There Once Were Some Sheep from Nantucket: Sheep in Colonial and Revolutionary America 7

Chapter 2. Embargos, Merinos, and the Flocks Grow: A New Republic of Wool 17

Chapter 3. "Utter Destruction to the Prosperity of That Section of the Union Whence I Come": Sheep and the Rift between North and South 34

Chapter 4. Shepherds Enslaved and Free: Sheep and Race 56

Chapter 5. Fido and the Fleece: Sheep, Dogs, and the Law 75

Chapter 6. Mary Had a Little Lamb and the West Had Big Burly Cowboys: Sheep, the Frontier, and Gender Bias 94

Chapter 7. Machines in the Pasture: Sheep and the Industrial Age 118

Chapter 8. Nothing Is Certain but Death and Taxes on Wool: Foreign Sheep and Internal Revenue   150

Chapter 9. "A Little Pure Wildness ... Both of Men and Sheep": Big Flocks, Bighorns, and Western Conservation   163

Chapter 10. Swimsuits, Soldiers, Polyester: The Decline of America's Sheep   174

Epilogue   191

List of Abbreviations   193

Notes   195

Index   239

Acknowledgments

Writing the acknowledgments for a book is simultaneously a happy task and one tinged with a bit of sadness. It is a great pleasure to have the opportunity to thank all the wonderful people who have supported me in my quest to write a book, but there is that nagging feeling that someone who deserves recognition will inadvertently be left out. My sincere apologies if that happens here.

When I worked at Stone Mountain Park in Stone Mountain, Georgia, Cindy Horton and Julie Furlong taught me a great deal about taking care of sheep, and I'll be ever grateful for that. I've had so many great colleagues at the Atlanta History Center, past and present, that I cannot mention them all. There are a few, however, that I'd like to single out. My position arose thanks to Sheffield Hale, president and CEO of the institution. When he took on those roles, he was disappointed that the Smith Farm on the History Center campus had no animals. He was determined to rectify this, and I was fortunate enough to be hired to take care of the sheep, goats, chickens, and turkeys brought in to populate the farm. My work has me reporting to Sarah Roberts, vice president of gardens and living collections, and I've profited immensely from her guidance. Tiffanny Jones, director of horticulture, inspires me with her professionalism and friendship. Michael Rose, the History Center's curator of decorative arts and special collections, encouraged me in this project; I owe him thanks for that and for putting me in touch with people who could help turn the book in my head into a reality.

If I could grant a wish to anyone writing a book, it would be that they be blessed with support as unwavering as I received from the University of Georgia Press. J. Patrick Allen was my editor in the early stages of the writing, with Nathaniel Holly sliding into that role later on. Both were a pleasure to work with and showed impressive toleration for my questions and missteps. As the book neared completion, Lea Johnson did a superb job as production editor. Chris Dodge was copyeditor, and because of him you think I write better than I do. Thanks to Ben Shaw for his indexing services.

In his insightful book *Our Lives, Our Fortunes, and Our Sacred Honor*, the late Richard Beeman, whom I never had the fortune to meet, wrote a tidbit on how the First Continental Congress's "Articles of Association" included a provision on the need for sheep improvement. That prompted me to further investigate the significance of sheep in the United States, and I wish I could personally thank Professor Beeman for the inspiration.

For help accessing information I wish to express gratitude to several of my colleagues at the Atlanta History Center, most especially to Jena Jones, manager of reprographic services, who worked her magic to turn the illustrations I wished to use into quality images suitable for publication. Selena Lim, designer and brand strategist, skillfully drew figure 10.3. Also helpful at my wonderful institution were Staci Catron, Cherokee Garden library director; Paul Crater, vice president of collections and research services; Ryan Nix Glenn, vice president of marketing and brand experience; Erica Hague, collections manager; Gordon L. Jones, military historian; Helen Matthews, librarian; and Sue Verhoef, director of oral history and genealogy.

Others I wish to thank are Zachary Bannor, data engineering manager, Condé Nast; Adam Berenbak, archivist at the National Archives and Records Administration; Warren Carver, natural resource specialist, U.S. Forest Service; Dru Cosner, site manager, Brasstown Bald Recreation Area; Christopher P. Donnellan, senior vice president of business affairs, Condé Nast; Thomas J. Forrest, librarian, Georgia Institute of Technology; Linda P. Gross, reference librarian, Hagley Museum and Library; Libby Hilliard, collections information and rights assistant, Crystal Bridges Museum of American Art; Renee Zuckerman Knapik, vice president of marketing and PR, InMocean; Lesley Martin, reference librarian, Chicago History Museum Research Center; Jim Orr, image services specialist, archives and library, Henry Ford Museum of American Innovation; Heather Richmond, archivist, State Historical Society of Missouri; Whitney Scullawl, collections assistant, Crystal Bridges Museum; Ryan Tolnay, formerly reference assistant at Mann Library, Cornell University; and Peregrine Wolff, wildlife veterinarian, Nevada Department of Wildlife.

Some of the information included in the first two chapters of this book I previously reported in an article published in the *Journal of the American Revolution*. I thank the editor of that publication, Don N. Hagist, for his immediate interest when I approached him with the topic and for his help bringing the article to fruition.

It has dawned on me repeatedly how indebted I am to the people and institutions who in recent years have digitized materials so that these may be accessed through the internet. I would not have had the time to go to an academic library to look up all the sources I ultimately ended up using to write this book. Furthermore, much of my research was done during the Covid pandemic when physical access to libraries was reduced or eliminated. Big thanks to everyone who has made it possible for research to be done far more conveniently and simply than could have been imagined just a few years ago.

Even with all the helpful advice from so many smart people, I'm still responsible for any errors in fact or judgment that appear in this book. I can't even blame my sheep for them.

Preface

Dana's Invisible Ancestors

> In the middle States, I have long been of opinion that we kept on our farms, too many black cattle, and too few sheep, and that a valuable revolution would be found in a reduction of the former, and augmentation of the latter.
> —JAMES MADISON to Robert R. Livingston, December 9, 1809

Rising to 4,784 feet above sea level, Brasstown Bald is the highest point in Georgia. Occasionally when the skies are unusually clear you can see Atlanta, sprawling a hundred miles to the southwest. Most days, however, it is too hazy atop the mountain for such a distant view, but an enchanting prospect will appear closer at hand—a panoramic series of peaks and valleys, outlined in that misty purplish profile so characteristic of Appalachian landscapes.

To visit Brasstown Bald, you drive along Georgia State Route 180, a two-lane mountain road—winding, woody, and especially scenic, lovely enough that you might wonder if you've ever driven another highway as beautiful. But you can't motor all the way to the top. The U.S. Forest Service maintains this area; when you arrive at its little booth you pay a fee and are instructed to pull into an adjacent parking lot. From there you can either hike to the summit or take the tram. If you take the tram, the driver provides interpretation to help you appreciate this short final leg of your journey.

At the peak of the mountain sits the visitor center. Like similar facilities in scores of parks across the land, this building has exhibits and displays to tell you all about the picturesque landscape that envelops you. You learn that, due to the altitude, the climate is more like New England than the lower elevations of Georgia—the highest temperature ever recorded at the summit was just 84°F. That is quite different from nearby Atlanta, where July days over 90°F are routine. If you visit on a spring day when T-shirts are worn at those lower elevations, you will be happy to have remembered to bring a jacket and to be wearing wool socks rather than cotton ones. As you move through the exhibits in the visitor center, you come to a section that focuses on the lives of the first European Americans to live in the area. In a panel titled "Money and Commerce," text states that the early settlers generally had very little money—understandable since the corn they grew sold for fifty cents a bushel. But by converting the grain to whiskey, the settlers created a product worth $1.50 a bushel, plenty of incentive to set up stills. The other staples of the settler economy are then enumerated. "Cattle, hides, honey, fruits, and nuts were bartered for essentials such as salt, gun powder, lead shot, shoes, and clothes."

Next to this panel is a very old black and white photographic image (see fig. 0.1). A mountain man stands tall and erect at center right. His shirt, coveralls, dusty shoes, and broad-brimmed hat give evidence that he is precisely that class of rural laboring man that the sign's words describe. He has a full beard, of course—after all, he is a mountain man! And what is that flat, round item he holds with both hands? Several livestock appear directly in front of him, and their mouths are pressed against the soil—but there isn't any apparent grass or other vegetation at this spot. Since it is bare ground, obviously these animals are not grazing. Clearly the item this mountain man holds is a plate with some corn. He is using a handful or so of his precious kernels to supplement his livestock's diet. One of them has come almost to his feet, head down, looking for a treat. It wears a cowbell that looks comically large, almost a third as big as its head.

But, in spite of its cowbell, this animal is not a cow. It is a sheep, as are all the other livestock in the photograph.

The display panel's text states that cattle were among those things mountain people bartered for salt and gunpowder. But there is not one word about sheep, the animals in the picture, on whose backs grew wool used for warm clothing, like the socks you're glad you wore here. It is almost as if the sheep are invisible, ghosts of a forgotten American past.

It is easy to believe that the United States was a nation of cowboys, not shepherds. You've seen cowboys in countless western movies and television shows. As a child you may have pretended to be a cowboy or a cowgirl. Perhaps, if you were raised a Christian, your childhood included assuming the role of a shepherd in December, when for the church Christmas pageant you wore a scratchy robe, held a mock staff, and were instructed to reverently

FIG. 0.1. Georgia mountain man and his sheep, undated. Photo is displayed in the USFS visitor center at Brasstown Bald, Georgia.

look at a makeshift manger that held a doll representing the baby Jesus. But that was something you only did once or twice; it wasn't a common occurrence like your cowboy and cowgirl games. It is likely that based on your experience and beliefs, your impression is that the United States was predominately a cattle nation, not a sheep nation.

That impression is wrong.

The first time I saw that photograph at the Brasstown Bald visitor center, I stared at the sheep with the big cowbell and exclaimed to myself, "Wow, she looks so much like Dana!" Dana was a favorite sheep of mine, a sweet ewe I had the pleasure of caring for in my job at the Atlanta History Center. Among the many treasures of our thirty-three-acre campus is the Smith Farm, which provides an authentic antebellum experience complete with the main farmhouse, a slave cabin, a blacksmith shop, a smokehouse, and other buildings, along with gardens and fields where we grow cotton, vegetables, and herbs—all this to better interpret life in the Atlanta area at the dawn of the Civil War. And, of course, there is the barn and its adjacent pen, home to our Gulf Coast sheep and Angora goats.

Dana was quite the little ovine character, particularly since she was so vocal. Next to the barnyard lies an abandoned rock quarry that several decades ago was converted to a garden of native plants. If I was down in the

Preface

quarry—near the barn but out of view—I could call out "Dana!" and have her respond with an extremely loud "Baaaaaaaa!" as she looked around for me. Visitors to the Smith Farm clearly get a lot of delight from hearing a sheep clearly call what they are stereotyped to say.

But many visitors are surprised just to see the sheep. "There were sheep in the South back then?" they inquire, never having contemplated such a thing. They are astonished when I reply that in 1860 there were over a half a million sheep in Georgia alone. (Maybe the one that looks so much like Dana in the photo with the mountain man was actually an ancestor of hers.) When I broaden the scope to include the entire nation, visitors are startled to learn that at the start of the Civil War there were over 21.5 million sheep in the United States. (As we will see in chapter 3, the antebellum South did not have enough sheep to fill a critical need, and this was a factor in the worsening relationship between the North and the South leading to war.) By 1880, sheep were more abundant in the United States than cattle—not what we would expect based on the popular image of late nineteenth-century America.[1]

For a time in the antebellum years, sheep even outnumbered people. In 1840 the United States had over nineteen million sheep and a human population of barely over seventeen million. The American tourist of today returning from a vacation in Australia or New Zealand and exclaiming to his friends and coworkers, "There are more sheep there than people!" is probably unaware that the same was once true of the United States.

Somewhere between looking at that photograph at Brasstown Bald and telling visitors to the Atlanta History Center that, yes, there were sheep in antebellum Georgia, I got the idea for this book. My research has caused me to follow a number of threads—threads figuratively of wool, if not literally—as they led me to discover people and places that make the saga of sheep in America so intriguing. From the time that sheep were brought to the New World by Europeans up until the mid-twentieth-century invention of polyester, sheep were frequently on the minds of many Americans, not just the people who maintained flocks in the field but also those who conducted the nation's business and those who managed the government in Washington. *American Sheep: A Cultural History* is their story. It is an attempt to reconnect with our heritage—and with Dana's ancestors.

AMERICAN SHEEP

INTRODUCTION

"The Protection of His Naked Body against the Inclement Elements"

> The United States of America, particularly those which lay to the north of the Chesapeake, appear to me to possess advantages in the breeding of sheep which are unequalled by any part of Europe which I have seen.
> —ROBERT R. LIVINGSTON, *Essay on Sheep*

He is not well remembered in our time, but in the early days of the United States, Robert R. Livingston was a famous statesman. A typical paragraph summarizing his resume highlights three accomplishments. First, along with John Adams, Benjamin Franklin, Thomas Jefferson, and Roger Sherman, he was a member of the Committee of Five that drafted the Declaration of Independence. Second, as chancellor of New York, he swore in George Washington as the first president of the United States. And third, as U.S. minister to France, Livingston, along with James Monroe, arranged the Louisiana Purchase.[1]

Livingston would probably wish that a fourth biographical detail about him was also familiar. He was the acclaimed author of *Essay on Sheep*, published in 1809.[2] The title is a bit misleading; at 186 pages it is more a small book than an essay. Livingston was then sixty-three, and his career in public service was behind him, giving him leisure to devote to his favorite pastime: expanding and improving America's flocks of sheep. "You see I am riding my hobby," he good-naturedly apologized in a letter to James Madison, after writing a long paragraph about sheep and their wool. "I must take care he does not run you down."[3] Had they been communicating face-to-face, Madison would likely have waved off any misgivings Livingston had over talking so much about his flock, since Madison was also quite interested in sheep. Most of the prominent people of the founding generation were.

1

Soon after its publication, a very favorable review of *Essay on Sheep* appeared in the *Medical Repository*, a widely read quarterly journal of medicine and natural sciences. The first two paragraphs of the review vividly express the high esteem sheep were held in by the revolutionary generation:

> In a cold country, wool is always a subject of importance to the inhabitants. In the eastern and northern states of the American Union it is peculiarly so. For more than six months of the year it may be considered as forming the chief material of clothing; and during the other six it enters largely into the covering of man. In such a climate, wool and the garments made from it, will always be in demand.
>
> The quantity of this important article, which the wants of the people require, will be constantly increasing. It will be called for as the human race multiplies. Every babe that is brought into the world is a new customer to the woolen draper. And the proudest as well as the meanest of mortals, derives the protection of his naked body against the inclement elements, from the sheep.[4]

In our time, when so much winter outerwear is made of synthetic fabrics, it is easy to forget the bygone days when the wool harvested off the backs of sheep stood between our ancestors and freezing temperatures. Wool had two big advantages over pelts from fur-bearing mammals, the only other pre-synthetic option for staying warm. One was that, like any domestic animal, sheep could be selectively bred to produce desired fleece characteristics, whether it be for wool fibers coarse or fine, long or short. (Providing instruction on how to do this was one of Livingston's goals.) Secondly, wool is harvested without killing its source. After shearing, a sheep is released to scamper back to its flock, where it begins the cycle of hair regrowth until the next clipping. Wool is thus a renewable resource. Fur bearers, on the other hand, must be killed to secure their pelts. This creates a resource shortage when the number of beavers, otters, or other mammals trapped and hunted outpaces their reproductive abilities, as happened in Europe and later in America. About a century before the American Revolution, the fur trade in Britain's Atlantic colonies was already in serious decline due to overharvesting.[5]

The review of *Essay on Sheep* remarked that wool was essential not only to the United States but also to residents of all climates with chilly winters. This would include the northern European ancestral lands of the colonists who settled the eastern coast of North America. It is thus worth noting that Florence, Italy, often called the birthplace of the Renaissance, became a prosperous metropolis largely because of its importance in the wool industry. Beginning in the thirteenth century, bales of wool were brought from English monasteries to Florence, there to be washed in the river Arno, then combed, spun, woven, and dyed into the cloth necessary for humans to thrive in climates like that of England.[6] With their naked bodies protected against the inclement elements by the sheep—as the *Medical Repository* had put it—Renaissance Europeans thrived, allowing them to make advances in arts, sciences, and commerce.

Wool also enabled them to embark on oceanic voyages. Of course, the ability of the Renaissance and post-Renaissance Europeans to do this typically was to the detriment of those already occupying those far-off places.[7] When English colonists stepped off their ships onto the lands of eastern North America, they were wearing wool. European cultures even began to consider their wearing of wool as proof of their superiority over the non-wool-covered cultures. "You must learn how to obtain the necessaries of life in plenty," President George Washington lectured the Cherokee Nation in 1796. "The most essential are food and clothing.... Some among you already experience the advantages of keeping cattle and hogs.... To them add sheep, and they will give you clothing as well as food."[8] Not having wool prior to European contact, indigenous peoples like the Cherokees kept warm by draping over themselves a cape-like garment called a "matchcoat," made of the skins of bison, beaver, or muskrats.[9]

One other characteristic of wool is extremely significant because it eventually led to a change in the perception of sheep by many Americans. To emphasize this characteristic, let us examine the painting *Winter Scene in Brooklyn*, painted by Francis Guy in 1820, just eleven years after Livingston's *Essay on Sheep* appeared (see fig. 0.2).[10] Since horses were critical to nineteenth-century Americans, it is no surprise that Guy's painting is dotted with equines, one or two of which bear riders and two of which pull a cart and a sleigh.[11] Three cows occupy the center of the picture. There are also two swine, six dogs, and a flock of chickens. A bevy of domestic animals are thus illustrated—but there are no sheep.

Looking at the painting another way, however, there are many sheep. Over two dozen human figures populate the painting, depicting winter in Brooklyn, New York, with snow and ice on the ground, and it is 1820, well over a century before the development of warm synthetic fibers. So the painting depicts people protected from cold by coats, trousers, dresses, and other clothing made of wool.

FIG. 0.2. Francis Guy, *Winter Scene in Brooklyn*, 1820, oil on canvas, Crystal Bridges Museum of American Art, Bentonville, Arkansas. Note that the early nineteenth-century urban landscape is populated with people, horses, cows, pigs, dogs, and chickens—but no sheep.

Why all that wool without a sheep in sight? I noted that sheep are not killed to harvest their fleece. Some other domestic animals also produce commodities that can be taken without slaughtering its provider, such as milk from dairy cows or eggs from poultry. But milk and eggs will spoil. Accordingly, before widespread use of refrigeration and swift, reliable transportation, the produce of dairy cows and chickens needed to be harvested close to where the people intending to consume them lived, thus the common presence of cows and barnyard fowl in urban America until well into the twentieth century.[12]

Not so with wool. Since it is not perishable it can be transported at leisure. It can even be stockpiled by those waiting for better market conditions. As such, it was never necessary for sheep to be kept close to the people who would eventually wear wool clothing or sleep under wool blankets. While there were some sheep in American cities, their presence there wasn't necessary in terms of a wool harvest. Sometimes, as we will see in chapter 8, it was simply a matter of them serving as living lawn mowers. More importantly, as argued in chapter 6, late nineteenth-century and early twentieth-century urban Americans, being less familiar with sheep in their everyday lives than with other domestic stock, were prone to adopt a view of sheep considerably less favorable than the one Robert Livingston held.

One further aspect of Livingston's essay deserves note: as much as he credited humans for domesticating sheep, he reciprocally credited sheep for civilizing humans. As he put it:

> That we are not at the moment fierce, savage, and brutal, little superior to the beasts that roam in the wilderness . . . is probably owing to the domestication of . . . animals, and, first of all, to that of sheep. To them we are also indebted for some of the most pleasing, as well as for the most important and useful arts. The cradle of music and poetry was rocked by the shepherds of Arcadia; while the spindle and the distaff, the wheel and the loom, originated in the domestication of sheep. This little animal, then, in losing its own wild nature, has not only converted the savage into the man, but has led him from one state of civilization to another; the fierce hunter it has changed into the mild shepherd, and the untutored shepherd into the more polished manufacturer.[13]

As we will see throughout this book, there is a bit of irony in praising sheep for removing from humans that which is "fierce, savage, and brutal," when the need for wool was a critical factor in American wars, from the Revolution of Livingston's world right on through the Korean conflict of the 1950s. For much of American history, the shearing of gentle sheep was the first stage in providing clothing for armed forces. The very existence of armies shows that while Livingston may have thought that sheep civilized humans into peaceful music- and poetry-loving shepherds of Arcadia, they were not entirely successful. America's acceptance of slavery was likewise savage and brutal; we will see in chapter 3 the significance of sheep to the practice.

To properly orient the story, however, we must jump back to examine the circumstances of American sheep beginning nearly two hundred years before Livingston wrote his celebrated book.

CHAPTER I

There Once Were Some Sheep from Nantucket
Sheep in Colonial and Revolutionary America

We lost many of our small Stocks of Sheep by the British Plunderers.
—GEORGE READ, letter to George Washington, February 5, 1778

Although the east coast of North America was colonized largely by English-speaking people, a man best known for his account of those lands was French. J. Hector St. John de Crèvecoeur was born Michel-Guillaume Jean de Crèvecoeur in Normandy in 1835. During his time in America, however, he took the Anglicized moniker of J. Hector St. John.[1]

Crèvecoeur came to the New World during the Seven Years' War, serving in the conflict as a cartographer for the French army. Following France's defeat in the Battle of Quebec in 1759, Crèvecoeur decided to stay in North America. After working as a surveyor and mapmaker, in 1770 he married and settled on a farm in Orange County, New York. There he and his wife began a family, and he began to write his observations on life in the early revolutionary years. These writings were gathered in *Letters from an American Farmer*, first published in 1782.[2]

While the title of Crèvecoeur's work highlights his role as a farmer, *Letters* comprises more than simply local agricultural observations. The French immigrant wrote about all things he found notable in his adopted land. His third letter, "What Is an American?," has been especially dissected by scholars as a rich source of understanding American identity of the time.[3]

In *Letters*, Crèvecoeur described the Massachusetts islands of Nantucket and Martha's Vineyard. Praising Nantucket's residents, he noted that in order to prosper, the Europeans settling there had to overcome the sparseness of the insular landscape. He called the island a "sandy spot" where nature had "refused those advantages" of arable land and trees. "There scarcely grew a shrub to announce, by the budding of its leaves, the arrival of the spring," Crèvecoeur lamented.[4]

As an island, Nantucket had the ocean to provide economic opportunity, and Crèvecoeur reported that two thousand seamen were constantly employed. But it was not only the sea that sustained the residents. While the interior of the island was unsuitable as cropland, the colonists found it valuable as rangeland for the scores of livestock they imported. There were on Nantucket, the Frenchman estimated, two hundred horses and five hundred cattle. Those numbers were dwarfed, however, by the island's sheep population—Crèvecoeur wrote that there were fifteen thousand of them.[5]

Truly impressive to Crèvecoeur was the manner in which Nantucket became a sheep mecca. Tracing local history, he reported that in 1671 twenty-seven proprietors gained authority over the island. Realizing it was too barren for cultivation, the proprietors agreed that after each of them staked out a forty-acre home lot, they would hold the rest of the land in common as sheep pasturage. But even though the land would be held jointly, each proprietor would have claim to his own sheep. The proprietors determined that each of them could keep a maximum flock of 560 animals, giving Nantucket a total population of 15,120 sheep. Crèvecoeur described this arrangement as "an ideal, though real title to some unknown piece of land" and notes that these holdings were appropriately called "sheep pasture titles."[6]

The abundance of sheep was a boon to the Nantucket citizens, Crèvecoeur declared: "They abound in wool; part of this they export, and the rest is spun by their industrious wives, and converted into substantial garments." But if the people of Nantucket were awash in wool, their neighbors on Martha's Vineyard were even more blessed—Crèvecoeur reported that this island boasted twenty thousand sheep, five thousand more than the Nantucket flock.[7]

This peaceful image Crèvecoeur painted of sheep grazing on the islands would be shattered before long when those insular sheep became a military target. To understand why this happened, we must first look back to the initial settlement of the Western Hemisphere by European people.

∽∞∞∽

When Christopher Columbus landed in the West Indies in 1492, a new era of animals and plants moving across the Atlantic Ocean in both directions commenced. This transfer of organisms across the globe is extremely significant toward understanding the environmental history of the earth we inhabit today.[8] Facilitated by the travels of humans, these movements of animals and plants were sometimes unintentional and unwelcome, as was the case with two rats of the genus *Rattus*. Originally confined to eastern and central Asia, these rodents spread as stowaways in caravans and ships, eventually coming to inhabit most of the globe. Disease and destruction of grain stores followed.[9]

Domestic sheep, on the other hand, being beneficial to humans, were purposefully transported across the seas. Already by 1439, over a half century before Columbus's voyage, sheep had been released on the Azores Islands in the mid-Atlantic, where they thrived and reproduced. Several European nations at the time sent ships to seek riches in West Africa; by introducing sheep on the Azores they had established a food source for ships returning home, when their provisions were diminishing.[10]

The Azores sheep would not be the last to be carried beyond Europe. In the 1500s, the Spanish brought sheep to many parts of the Western Hemisphere, including the lands that later became the southeastern United States. The Gulf Coast breed of sheep are the living descendants of these sixteenth-century introductions.[11]

The English too viewed sheep as an integral part of their colonial plans. Early in the seventeenth century they transported sheep to their nascent colonies of New England and the Chesapeake region.[12] Their efforts to establish domestic livestock in the North American colonies was not confined to sheep; cattle, swine, and horses also made the shipboard trip across the Atlantic. But while other livestock frequently thrived on their adopted continent, sheep often did not do well and sometimes perished. One difficulty complained of by the colonists was that, in the absence of the managed pastures of England, sheep had to forage in forests and thickets, where the abundance of tall, sometimes thorny vegetation snagged their wool and tore it off their backs.[13]

A far bigger problem was wolves. While all livestock were potential prey, sheep were particularly susceptible to wolf predation.[14] This would not have been such a setback if the colonists depended on sheep only for meat, other livestock and wild deer being available to supply their tables. But flock losses to wolves—and to a lesser extent bears—became a major concern since sheep were essential for their wool.

In the mid-seventeenth century, with England gripped by political turmoil and civil war, the New England colonists faced persistent disruptions in their trade with the mother country. Cut off from a reliable source of clothing, they grew ever more desperate for apparel, especially warm garments needed as protection from the harsh winters. More intensive management of sheep would be necessary if the colonists were to have sufficient stores of homegrown wool. They responded to the crisis in a number of ways. Bounties were offered, both for the heads of wolves and for the production of wool cloth. Efforts were undertaken to clear brush to provide improved pastures. In Massachusetts, laws prevented the slaughter of ewes, ewe lambs, or even any male sheep under two years old, so to insure that sheep were providing wool instead of dinner.[15]

The colonists did another thing. Repeating the experience of two centuries earlier in the Azores, they put sheep on islands, such as Nantucket and Martha's Vineyard. There, as we have seen, the flocks thrived. In large part, the successful establishment of sheep offshore was due to the absence of large predators.[16] But while islands were the best place to keep sheep safe from wolves, the colonists would later learn that they were the worst place to keep sheep safe from the Royal Navy.

Since their settlement of the Atlantic coast, the colonists had not agreed with every law or policy imposed on them by England, but the 1765 Stamp Act is commonly cited as the crucible leading to the American Revolution.[17] The first effort by Parliament to directly collect revenue from the colonists, the Stamp Act placed a tax on legal documents and published materials. Opposition was immediate and widespread.[18]

Stationed in London as an agent of the Pennsylvania Assembly, Benjamin Franklin was called before the House of Commons in February 1766 to testify as to the extent of colonial objection to the Stamp Act. It turned into a broader examination of England's relationship with its colonies. Franklin was peppered with inquiries, some rather hostile. "Do you think it right that America should be protected by this country, and pay no part of the expense?" a questioner demanded.[19] "Are not the lower rank of people more at their ease in America than in England?" another asked of Franklin, pointedly and probably rhetorically.[20]

Almost midway through the questioning, Franklin was asked if cloth manufactured in England was "absolutely necessary" to the colonists. He denied that it was, but the intention of the question was clearly to turn the interrogation into an exchange on the state of sheep in the colonies. Could the people possibly find enough wool in North America? Franklin confidently asserted that they could, declaring that the colonists had taken steps to increase their wool. "They entered into general combinations to eat no more lamb, and very few lambs were killed last year," he reported. "This course persisted in, will soon make a prodigious difference in the quantity of wool."[21]

Obviously, the House of Commons was skeptical about that, as they posed several follow-up questions. "Does not the severity of the winter, in the Northern Colonies, occasion the wool to be of bad quality?" Franklin was asked. He disagreed, saying the quality was just fine. Then came a question suggesting that as bad as northern wool was, that from the South was even worse: Wasn't the wool from sheep in Virginia very coarse, more like hair than wool? Franklin admitted he was not sure about that, but, if it was true, so what? Southern winters are milder than northern ones, he declared, so colonists in Virginia, the Carolinas, and Georgia did not need as much wool as their northern brethren. They could wear linen and cotton most of the year, he asserted. There was one final inquiry on the subject: Franklin was asked if the climate in the northern colonies was so severe that sheep had to be provided fodder all winter long. Franklin admitted that in some places this was so.[22]

Privately Franklin had admitted in a letter written in the fall of 1764, "As to Clothing ourselves with our own Wool, 'tis impossible. Our Sheep have such small Fleeces, that the Wool of all the Mutton we eat will not supply us with Stockings."[23] Franklin stood before Parliament only eighteen months later; if he was right in his private letter about the paucity of raw wool in the colonies, a year and a half was not enough time for America's flocks of sheep to have increased sufficiently to relieve such a shortage. So he was being evasive and unjustifiably optimistic about colonial wool at his London questioning.

His appearance at the House of Commons was not the only time Franklin exaggerated about American wool, but the other instance where he did this was in jest. In May 1765 he wrote a satirical article using the pseudonym "A Traveller." In that essay, Franklin mocked those who said that the sheep in America had so little wool they could not even provide one pair of stockings a year to each colonist (what he himself privately admitted). A Traveller insisted, "The very Tails of the American Sheep are so laden with Wool, that each has a Car or Wagon on four little Wheels to support and keep it from trailing on the Ground."[24] During his 1766 interrogation by Parliament, Franklin probably sensed it was not a good time for such levity.

Franklin's London interview did nothing to alleviate the worsening relationship between Britain and the colonies, and eight years later in September 1774 the First Continental Congress assembled to determine a unified American action. It was a distinguished group that took their seats in Philadelphia, including George Washington and John Adams, later to become the first two presidents of the United States; John Jay, later the first chief justice of the Supreme Court; and such celebrated patriots as Samuel Adams, Patrick Henry, and Richard Henry Lee.[25] The assembly quickly decided that the most effective means of expressing their resolve was to ratify an agreement that the colonies would no longer import goods from Britain or its possessions nor export goods to them. This policy was promulgated in a declaration that has come to be known as the Articles of Association.

If the colonies planned to give up all British goods, that meant no more woolen cloth or clothing arriving on ships in colonial harbors. What would the Americans do to alleviate that foreign dependency? The seventh article addressed this important matter. The representatives resolved:

> We will use our utmost endeavors to improve the breed of sheep, and increase their number to the greatest extent; and to that end, we will kill them as seldom as may be, especially those of the most profitable kind; nor will we export any to the West Indies or elsewhere; and those of us, who are or may become overstocked with, or can conveniently spare any sheep, will dispose of them to our neighbors, especially to the poorer sort, on moderate terms.[26]

By "improving" the breed of sheep, the representatives meant the selective breeding of flocks to develop animals with more wool and of better quality. As discussed in the next chapter, sheep improvement became something of an obsession for some early Americans.

The first part of article 7, a promise not to kill sheep for food but instead work diligently to increase and improve the flocks, restates what Franklin had told the House of Commons in 1766. The long sentence then moves into a declaration of republican virtue. Much has been written about the great importance the founding generation placed on the necessity of virtue to the success of free government. Article 7 clearly expresses that public virtue, a quality defined by historian Gordon Wood as "the willingness of the individual to sacrifice his private interests for the good of the community."²⁷ The delegates to the First Continental Congress, many of whom were men of some means, were asserting on the record that it would be improper to take advantage of the ongoing wool shortage to get top dollar for surplus sheep; in tough times patriotism commanded that extra animals be dispersed "on moderate terms."

British loyalists living in the colonies were aghast at the actions of the First Continental Congress. Prominent among them was Samuel Seabury, the first American Episcopal bishop.²⁸ In a ninety-page pamphlet written under the pseudonym "A. W. Farmer," he blasted the Articles of Association, and the seventh article did not escape his ire. The Reverend Seabury sneered,

> [If wool is not imported,] the first winter after our English goods are consumed, we shall be starving with cold. *Kill* your sheep ever so *sparingly*, keep every [castrated male sheep] as well as ewe, to increase the number and improve the breed of sheep; make every other mode of farming subservient to the raising of sheep, and the requisite quantity of wool to clothe the inhabitants of this continent, will not be obtained in twenty years: if they increase only as they have done, not in fifty (emphasis in original).²⁹

Seabury was responding not only to the Articles of Association but also to a pamphlet written by a brash, pro-independence teenager—Alexander Hamilton. Late in 1774, Hamilton endorsed article 7 and even went a little further than the First Continental Congress delegates. Writing as "A Friend to America," he confidently declared:

> We can live without trade of any kind [with Britain]. Food and clothing we have within ourselves. Our climate produces cotton, wool, flax and hemp, which, with proper cultivation would furnish us with summer apparel in abundance. . . . We have sheep, which, with due care in improving and increasing them, would soon yield a sufficiency of wool. The large quantity of skins, we have among us, would never let us want a warm and comfortable suit.³⁰

Hamilton was being unjustifiably boastful. Where the "quantity of skins" was concerned, the America of Hamilton's time was already short of furs from wild beasts due to overharvesting. But as to the topic of interest here, sheep's wool, it would play out that Reverend Seabury had a far better understanding of America's limits than did his young rival. The British had an advantage they would press when their quarrel with the colonies turned into full-scale war.

~~~

John and Abigail Adams kept sheep at their family farm in Quincy, Massachusetts. John enjoyed the time he was able to engage in agriculture, but he was often away from home serving his country in official capacities.[31] In his absence, Abigail would write letters, filling her husband in on the details of their farm, especially the expenses. In one letter, written late in winter, Abigail obviously looked forward to spring, when money could be saved because their sheep could once again graze. "Tis a sad expensive thing to have to feed sheep 4 months with corn and English Hay," she lamented, showing that Benjamin Franklin was correct when he acknowledged in Parliament that sheep in the northern states had to be provisioned during the winter.[32]

But there was worse news about sheep reported in another of Abigail's letters. When John Adams traveled to Paris to serve as revolutionary America's envoy to France, he took his young son John Quincy with him. Abigail wrote to both of them frequently. The letter she sent to John Adams dated September 29, 1778, was full of grim details about American setbacks in the war and equally unpleasant accounts of the family finances.[33] By contrast, the correspondence Mrs. Adams sent to John Quincy that same date was mostly the expression of love and concern to be expected in a mother's letter to her eleven-year-old son whom she had not seen in months. She inquired about his health, asked about his progress learning French, and declared, "The next opportunity you have for writing you must not forget your Grandmamma."[34] But Abigail added one notable detail about the war not in the note she sent that same date to her husband. Young John Quincy read in his mother's letter that earlier in the month the British—in an operation now known as Grey's Raid—had plundered nine thousand sheep from Martha's Vineyard.[35] The British targeted coastal and insular sheep in other places along the Atlantic coast as well.[36] Given the American need for wool, these were military actions analogous to bombing oil fields or refineries in modern wars.

All that sheep plundering had the effect that British forces desired. Early in 1778, even before Grey's Raid, General George Washington dashed off a grim letter from his Valley Forge headquarters to the American Board of

War detailing how meager supplies had become for his troops that winter. One shortage particularly concerned him:

> As to Blankets, I really do not know what will be done. Our situation in this instance is peculiarly distressing. I suppose that not less than from 3 to 4000 are now wanted in Camp—Our Sick want—Our Unfortunate men in captivity want.[37]

Washington did not need to connect the dots by adding "blankets are made of wool, which comes from sheep, of which we have pitifully small flocks, which are under assault from the enemy, and the war has cut off imports so we can't get blankets from overseas." The men on the Board of War knew all this, but it is a point that can be missed by modern readers of the letter. The failure to adequately equip the American troops is often seen as a problem with the supply system—a shortage of funds to purchase provisions plus a dearth of wagons and suitable roads.[38] But, when it came to wool for clothing and blankets, there was a more basic problem—America just did not have enough sheep. Nor did it have sufficient manufacturing facilities to turn whatever domestic product existed into enough woolens for all the troops, let alone for the civilians. On top of all their other wartime difficulties, the colonials faced a severe wool crisis.

Yet the Americans prevailed. In his 1783 farewell address to the army, Washington acknowledged that the revolutionary cause could easily have suffered a crushing defeat. "The unparalleled perseverance of the Armies of the United States, through almost every possible suffering and discouragement, for the space of eight long years, was little short of a standing Miracle," he reminded his troops.[39] As we have seen, significant among the suffering and discouragement that could have led to an American debacle was a shortage of sheep. Reverend Seabury's prediction that Americans would be "starving with cold" was a real possibility.

There is a notable postscript. While much of Great Britain's economic prominence in the Industrial Revolution was due to the nation's textile mills churning out wool garments and blankets, the British themselves had to import the raw material. Their flocks of sheep simply were insufficient to provide those factories with all the wool needed. Robert Livingston pointed out in his *Essay on Sheep* that most British ovines produced middling-grade fiber. Elaborating, he asserted that wool from the island's flocks was not "of sufficient fineness for broadcloths of the first, second, and third qualities;

these are all made from Spanish wool . . . without admixture. Of this wool near seven millions of pounds are annually imported into Britain."[40] Thus, the process leading to wool clothing manufactured in England, and worn by revolutionary-era Americans, often began with sheep herded and pastured on the Iberian Peninsula.

This meant that not long after the Revolution, the British, like the Americans they had recently battled, would themselves face a sheep crisis as continental Europe became politically unstable following the French Revolution and later the spread of Napoleon's armies.[41] Just a year and a half after the Storming of the Bastille, the British politician and agriculturist Sir John Sinclair sounded the alarm that his nation had to pay better attention to the quality of its own sheep. He lamented that Britain was "obliged to import considerable quantities of clothing wool from another kingdom," and he urged his countrymen, "We cannot therefore too soon endeavor to remedy what . . . must be considered as an evil before it takes too deep a root, and becomes more difficult to eradicate."[42] Sinclair sounded quite like a typical American statesman of twenty-five years earlier.

It thus occurred to both the British and the Americans that rather than depend on Spain for sheep, it would make sense to bring Spanish sheep to their respective homelands—or in Britain's case, with acreage at more of a premium, to take those livestock to other parts of the empire, such as Canada, Australia, and New Zealand.[43] Efforts in the United States to supplement American sheep with stock from Europe were widely supported, although, as we will see in the next chapter, there was some disagreement on whether sheep improvement was a function for government or for private citizens.

CHAPTER 2

Embargos, Merinos, and the Flocks Grow
A New Republic of Wool

> Amidst these restrictions on the intercourse of nations, it has been found expedient to begin a woolen manufacture at home.
> —ANONYMOUS, review of *Essay on Sheep*

Maryland congressman Joseph Nicholson's survey of potential non-British sources for wool did not look promising, he conceded when he shared his findings on the floor of the House of Representatives in February 1806. Nicholson told his colleagues that neither Germany nor Holland nor any other place on the European continent could provide goods in the amounts Americans consumed. "With coarse woolens," he unhappily concluded, "we are supplied altogether from Great Britain, and we cannot procure them elsewhere."[1]

As Nicholson spoke, Britain was at war with Napoleon's France. Because of the hostilities, British ships seized goods on American ships to keep them out of French hands. The British also boarded U.S. ships to impress American seamen believed to be deserters from the Royal Navy. Angered by these actions, which the United States considered an affront to its official position of neutrality between the combatants, Congress debated banning British imports in retaliation.[2]

Considering American dependence on British wool, Congressman Nicholson argued that it would be nonsensical to enact a complete embargo: "We shall be laughed at by Great Britain and all other European powers for adopting a system altogether impracticable, because we cannot adhere to it."[3]

Nicholson then floated an idea. While he had reported that the woolens worn by Americans of ordinary means could only be acquired from Britain, he hastened to add that there were other options for "those who moved in the higher walks of life." This being so, he suggested not banning the importation of all British woolens but only the more expensive ones.[4]

And that is precisely what happened. In April 1806 Congress passed an embargo prohibiting the import of several British goods. The list included a total ban on several items used for clothing—all articles of leather, silk, and hemp were disallowed. When it came to wool, however, the legislation only banned "woolen cloths whose invoice prices shall exceed five shillings sterling per square yard" and "woolen hosiery of all kinds." The act also prohibited the import of all "clothing ready-made."[5] Americans could still get from England cheap woolen cloth, not already assembled into garments. Only the well-off who had favored ready-to-wear clothing from England would have to change their consumption habits.

This half-hearted regulation of woolens was a stark acknowledgment that, three decades into its founding, America still was not self-sufficient when it came to sheep. It had been forty years since Benjamin Franklin had stood before Parliament and assured them that their North American colonies were taking strong steps to increase their wool. It had been thirty-two years since the Articles of Association pledged to increase and improve sheep—and this after banning import of *all* British woolens, not just the expensive kind. Clearly there was still work to do for the new nation to meet its goal of having enough American sheep. But the efforts to minimize dependence on European wool were about to kick into a higher gear, and, like the 1806 Embargo Act, the Napoleonic Wars had a lot to do with this. Even more critical to the upgrading and expanding of America's flocks was attentiveness to a process that humans had first discovered millennia earlier.

When a scientist proposes a revolutionary theory, objections might be lessened if the hypothesis is introduced with an appeal that some of its applications are already in use. So it was in 1859 when Charles Darwin published *The Origin of Species*. The book's first chapter was titled "Variation under Domestication." Here Darwin argued that a simplified version of his theory of evolution through natural selection took place whenever humans selectively bred domesticated animals to promote desired characteristics. The careful attention sheep farmers paid to their charges particularly impressed him. He noted that in Saxony the process was so meticulous that sheep were "placed on a table and . . . studied, like a picture by a connoisseur" ("this is done three times at intervals of months, and the sheep are each time marked and classed, so that the very best may ultimately be selected for breeding").⁶

As Darwin knew, knowledge that features of individual organisms could pass to successive generations—and the corresponding understanding that humans could partly control the process by deciding which sire to mate with a particular dam—stretched back to antiquity. Since his theory is in part a refutation of the Genesis origin story, it is notable that in *The Origin of Species* Darwin cited another portion of the same Old Testament book as factual—and once again sheep were involved. Just a few paragraphs after praising the Saxony shepherds, Darwin wrote: "From passages in Genesis, it is clear that the color of domestic animals was at that early period attended to."⁷ Clearly he was referring to Jacob and the spotted sheep.⁸ In this story, Jacob manipulates the breeding of sheep and goats, leading to a flock where spotted and speckled animals dominate.⁹ The tale of Jacob's sheep has been called "the first historical example of applied Darwinism."¹⁰ In other words, long before *The Origin of Species* appeared, long before the modern study of genetics, humans knew the principles of heredity. They just did not understand the mechanics behind it.

The founders of the United States of America—managing their flocks decades before Darwin published his master work—were a part of that venerable tradition. Some of the men who rebelled during the Revolution focused on the selective breeding of sheep for the remainder of their lives. While he served as president, George Washington left the care of his agricultural business, including at Mount Vernon, to his farm manager Anthony Whitting. In June 1793, early in his second term, Washington received a letter from Whitting detailing the results of the recent shearing. The president was quite irritated. His 568 sheep had, according to the correspondence, provided 1,457 pounds of wool—an average of a little over two and a half pounds per animal. Washington was certain that the yield should be much higher; he clearly felt that in his absence careful attention to ovine reproductive control had fallen by the wayside. When he was present in Virginia looking after his own flocks, Washington boasted, "I improved the breed of my Sheep so much by buying, & selecting the best formed, & most promising Rams & putting them to my best Ewes . . . that they averaged me . . . rather over than under five pounds of washed wool each."¹¹

By this point, Whitting no doubt realized that every letter from his boss was going to contain stern directives about sheep or a complaint that the flocks were not being well attended. Only a month earlier, Washington grumbled that while home the previous summer he saw rams with his ewes, but he was not sure if those rams were his or belonged to someone else. If the rams were property of one of his neighbors, he wrote in a letter to Whitting, that would be horrible. "I believe my Sheep are above mediocrity, when most others are below it."[12] In another letter dated just two weeks later, Washington elaborated on the fine points of sheep husbandry that he insisted be followed, ending his list of instructions: "In a word—I wish every possible care may be used to improve the breed of my Sheep; & to keep them in a thriving & healthy State.[13]

Washington was not alone in his attentiveness regarding upgrading the new nation's sheep; private efforts toward this end were common. But when Washington gave the first-ever State of the Union message, it led to a proposal by Treasury Secretary Alexander Hamilton that sheep improvement be instituted as a matter of government policy.

༄༅༅

"To be prepared for war is one of the most effectual means of preserving peace," President Washington insisted when he addressed Congress in January 1790. It was thus only fitting that to insure America's safety the infant nation should "promote such manufactories, as tend to render [the United States] independent on others, for essential, particularly for military supplies.[14]

A week later the House of Representatives passed a resolution instructing the secretary of the treasury to prepare a report consistent with Washington's recommendation, using almost identical language.[15] Neither Washington's speech nor the House resolution said anything specific about wool. But given the highlighting of military supplies by both the executive and the legislative branches, it is not surprising that when Secretary Hamilton presented his *Report on the Subject of Manufactures* to Congress, a discussion of wool was included. Hamilton had, after all, served as Washington's aide-de-camp during the Revolutionary War.[16] In the report, he wrote: "The extreme embarrassments of the United States during the late War, from an incapacity of supplying themselves, are still matter of keen recollection."[17] Hamilton remembered the shortage of blankets that Washington had lamented in his letter to the Board of War during the Revolution.

Historians have noted the influence of Adam Smith's *An Inquiry into the Nature and Causes of the Wealth of Nations* on Hamilton's *Report on*

Manufactures.[18] References to sheep and their wool figure prominently in Smith's treatise—it could scarcely be otherwise in an eighteenth-century volume on economic policy written by a resident of Great Britain. From his study of *The Wealth of Nations*, Hamilton would have encountered some odd minutiae about the wool trade. He would have read about an archaic unit of measure for wool called a "tod," which was equal to twenty-eight pounds of English wool. One tod in 1339, Hamilton would have further learned, sold for around ten shillings.[19]

More significantly, reading *The Wealth of Nations* would expose Hamilton to Smith's criticism of the protectionist wool laws of his homeland. Smith lamented that Great Britain's wool manufacturers had for many years convinced Parliament that the prosperity of the nation depended on a thriving wool industry. As a result of this successful lobbying, Parliament had passed some very harsh legislation to prevent exportations out of Britain. A law enacted in Queen Elizabeth's time punished a first offense exporting sheep by making the offender forfeit all his goods forever, imprisoning him for a year, and cutting off his left hand. A second offense led to the death penalty. Later, during the reign of Charles II, the law was expanded to ban the exportation not just of live sheep but also of their sheared wool. The severe penalties of the earlier law were also applied to the new act. "For the honor of the national humanity," Smith solemnly wrote, "it is to be hoped that neither of the statutes were ever executed."[20] Whether or not hands were cut off or offenders hung, however, the economic sanctions alone were effective deterrents.

Not content merely to ban exports of British wool, Parliament also prohibited its exportation from Ireland to any country other than England, and it permitted duty-free importation of wool from Spain.[21] The effects of these actions, Smith reported, had caused an artificial depression in the price of wool in Great Britain, such that its value at the time he wrote *The Wealth of Nations* was less than it had been in the fourteenth century.[22] Smith concluded that an absolute prohibition on the exportation of wool was not justified but that a significant tax on exports would be sound fiscal policy.[23] In other words, he did not reject the idea that the British government should have a role in regulating its vital wool industry; he simply saw Parliament's proper involvement as being less onerous and more responsive to England's entire economic health, not chained to the protectionist desires of a few.

Hamilton hoped to put some of Smith's ideas into practice in America, including those related to wool. While the title *Report on Manufactures* suggests finished products at the end of a factory line, Hamilton's consideration of wool mainly centered on government support for the source of the raw material. "Measures, which should tend to promote an abundant supply of wool, of good quality, would probably afford the most efficacious aid, that present circumstances permit," he wrote. It thus behooved the United States "to encourage the raising and improving the breed of sheep."[24] Hamilton proposed that the government reward those Americans raising superior sheep by paying premiums, which he defined as rewards for "some particular excellence or superiority, some extraordinary exertion or skill, and . . . dispensed only in a small number of cases." But while only a few sheep farmers would be awarded the premium, many shepherds, seeking to earn the monetary reward, would take care in the selective breeding of their sheep. Thus, Hamilton argued, the premium would be "a very economical mean of exciting the enterprise of a Whole Community."[25]

In essence, Hamilton was proposing that the U.S. government assume a duty we now associate with judges at a county fair, awarding a blue ribbon to the farmer whose livestock is deemed the best of any displayed. Hamilton's effort was for naught; Congress was unconvinced by his arguments on wool or any other commodity, and it shelved his report.[26] Had Congress implemented the sheep recommendation in *Report on Manufactures*, it would certainly have been interesting to see the specifics and the logistics involved in coordinating an eighteenth-century nationwide search to reward excellence in the field of raising superior sheep.

Where in the Constitution that Hamilton helped to write did he find justification for the U.S. government to offer the sheep premium he endorsed? He argued that this was authorized by the power granted Congress "to lay and collect Taxes . . . to . . . provide for the common Defense and General Welfare of the United States."[27] That is a rather broad reading of the general welfare clause, and Hamilton's expansive interpretation would be challenged by his frequent political rival.

Thomas Jefferson purchased books from the French immigrant bookseller J. Phillipe Reibelt, and the two corresponded regularly. Their letters were not confined to literary matters; they sometimes ventured into politics—which, given the era, meant that sheep were discussed. Early in Jefferson's second term as president, Reibelt sent him a somewhat breathless note, recounting how a number of European nations—among them Great Britain, Austria, Prussia, Holland, Sweden, and Denmark—were hard at work acquiring new stocks of sheep, the precious Merinos of Spain. Wouldn't it make sense for the United States to implement a similar policy?[28]

Jefferson courteously sent an immediate reply. Yes, importing sheep would be to America's benefit, he agreed. But a legislative bill to do so was quite out of the question, because "the general government (possesses) no powers but those enumerated in the constitution." So, regarding the sheep specifically, he wrote, "Congress could not, by our constitution give one dollar for all in Spain, because that kind of power has not been given them."[29] Even though the Constitution gave Congress authority "to regulate Commerce with foreign Nations, and among the several States," and even though sheep produced wool, an item of commerce sorely needed in the United States, for Jefferson this was not a sufficient nexus to permit Congress to purchase sheep from overseas.[30] Given his strict construction of the Constitution, it is unsurprising that Jefferson also said nothing about wool as a military necessity, which might arguably authorize him to act under his own executive power as commander in chief.[31]

Historians commonly note the disagreement between Hamilton and Jefferson on the constitutionality of a national bank, with Jefferson arguing for a strict construction of the Constitution that would disallow such an institution and Hamilton advocating a looser construction that would authorize the bank.[32] The two were no less at odds in their constitutional interpretation when it came to sheep. Hamilton declared that the general welfare clause enabled Congress to financially aid the development of wool manufactures; Jefferson instead believed that no provision gave Congress authority to acquire additional sheep necessary to produce the wool in the first place.

But by asserting that Congress could not act, Jefferson did not mean to imply that there was no way to bring sheep across the Atlantic. To him it simply was not a matter for government but instead for virtuous citizens, including himself, acting independently. As he put it to Reibelt: "It is probable that private exertions will transplant & spread [Merino sheep]." He continued proudly: "I have possessed the breed several years, and have been constantly distributing them in my neighborhood."[33] In this effort, Jefferson was part of an interconnected web that early in the nineteenth century joined a number of prominent Americans.

Among those onboard the *Benjamin Franklin* crossing the Atlantic Ocean in 1801 were two of the oddest of traveling companions: E. I. du Pont and Don Pedro. Du Pont was French and a human, while Don Pedro was Spanish and a sheep (see fig. 2.1). It was Don Pedro's first transoceanic passage, but du Pont was making the voyage to the United States for the second time. The year before, he had arrived in America and promptly concluded that the young nation needed more gunpowder and better sheep. Quickly back to France he went, seeking machinery and workers for a gunpowder factory and Merino rams to improve US flocks. Don Pedro was fortunate to have survived; he was one of four young rams du Pont procured but the only one still alive when the *Benjamin Franklin* sailed into port in Philadelphia that summer. He would not be the only purebred Merino in the country for long, however. During the next year, 1802, Robert Livingston, serving as ambassador to France, used his connections to procure a few of this prized breed for his New York farm. More significantly, David Humphreys, the ambassador to Spain, acquired nearly a hundred Merinos to stock his farm in Connecticut.[34]

The Merino breed has a lively history. For centuries the sheep of the Iberian Peninsula had been highly valued for their fine wool. In the late Middle Ages, Spain initiated efforts toward sheep improvement similar to U.S. aspirations centuries later. By the fifteenth century, all aspects of Spanish sheep management had come under control of a quasi-governmental association called the Mesta. Because of its affiliation with the Crown, all the Mesta's policies and practices regarding sheep had the force of law. Under this rigid scrutiny, the Merino breed was developed. Realizing how significant Merino wool was to its economy, for a long time Spain did all it could to maintain a monopoly by keeping all these prized sheep within its borders.[35]

By the late eighteenth century, however, Spain's influence in world affairs had waned; accordingly, as a matter of diplomacy, Merinos were sent to other European nations. As a stipulation of the 1795 Treaty of Basle, for example, Spain sent six thousand of these sheep to France, and it was out of this flock that Don Pedro came. Other Merinos were smuggled out of Spain, especially to Great Britain. As noted, a few also trickled into the United States. But these

FIG. 2.1. Don Pedro the Merino ram. From *Register of the Vermont Merino Sheep Breeders' Association*, vol. 2, 1883.

DON PEDRO.
Imported from France by M. Desselert, 1801.

emigrant flocks remained too small to compete effectively with Spain in the fine wool trade—until the Napoleonic Wars. Following the French invasion of Spain in 1809, perhaps two hundred thousand Merinos were driven into France; thousands also were shipped to England.[36] Thanks to William Jarvis, American consul in Lisbon, the United States also benefited. Jarvis took advantage of the wartime situation to acquire thirty-five hundred Merinos for export to his country.[37] Though Jefferson had written that if Merinos were to be procured from Spain it would have to be through private efforts, it was a government agent, given his commission by President Jefferson, whose actions helped put the Merino on a firm footing in America. The supply of Merinos remained scarce enough, however, to cause a speculative bubble. This led to exorbitant prices of three to five hundred dollars a head, with some rams selling for a thousand dollars. As such, Merinos were primarily an indulgence of well-off gentlemen farmers.

But then, in 1810 and 1811, around twenty thousand more Merinos were imported into the United States, and this increased supply reduced prices.[38] So did another factor. While some Merino keepers tried to maintain purebred flocks, more common was the practice of crossing Merinos with the so-called native sheep that had been the mainstay of American flocks prior to the Spanish importations.[39] The progeny resulting from such crosses produced wool of improved quality and quantity, obviously of higher importance to a young nation trying to grow more wool than any aristocratic notion of breed purity. By the fall of 1810, sheep with only half Merino ancestry were selling for between thirty and fifty dollars a head, bringing them within the reach of farmers of more modest means than men like Livingston and Humphreys.[40]

The high prices of Merinos before the later imports had irritated Thomas Jefferson, who charged that business and patriotism had become unacceptably intertwined.

What is a patriot? Judge Richard Peters of Philadelphia provided an ovine gloss to the word. In a lecture delivered in 1810, he noted that the first sheep of the Tunis breed to be imported into America were due to the efforts of the US consul at Tunis, William Eaton. "For this estimable proof of his patriotism," Peters applauded, "he merits the thanks of all who profit by its advantages."[41]

Peters was not the only one linking sheep improvement to patriotism. Livingston, in his *Essay on Sheep*, lavishly praised the Arlington long-wooled sheep, acclaiming the originator of the breed, George Washington Parke Custis. "I cannot omit this occasion," Livingston gushed, "to express the high opinion which I, in common with every other person, entertain of Mr. Custis's patriotism, and of his animated exertions for the improvement of this most important branch of our rural economy."[42]

Furthermore, when William Jarvis sent two of those Merinos he had procured in Lisbon to Thomas Jefferson, he offered the gift humbly, writing, "I hope you will allow me again to trespass on your goodness with a small present, which I trust, from your Patriotism, will not be unacceptable."[43]

Merriam-Webster defines "patriotism" simply as "love for or devotion to one's country." In Edward Gibbon's *The Decline and Fall of the Roman Empire*—a work familiar to many Americans of the founding era—there is a more elaborate description of the term. "Patriotism," Gibbon declares, "is derived from a strong sense of our own interest in the preservation and prosperity of the free government of which we are members."[44] That meaning of the word must have pleased America's founders, who took pride in creating what they considered the freest government possible. But as the above examples show, there was an application of the concept of patriotism known to America's founders that is forgotten today. In their time one was considered a patriot if one did one's part to increase or to improve the nation's stock of sheep. In doing so one was, in Gibbon's terms, aiding the preservation and prosperity of the United States.

But Thomas Jefferson insisted that "patriot" was not synonymous with "profit." Jefferson's friendship with James Madison led him to often candidly share his feelings with the younger Virginian.[45] In a letter written in 1810, when Madison was president and Jefferson the recently retired one, Jefferson fumed over the business practices of those selling Merinos:

> I have been so disgusted with the scandalous extortions lately practiced in the sale of these animals, and with the description of patriotism and praise to the sellers, as if the thousands of dollars apiece they have not been ashamed to receive were not reward enough. . . . No sentiment is more acknowledged in the family of Agriculturists than that the few who can afford it should incur the risk and expense of all new improvements, and give the benefit freely to the many of more restricted circumstances.[46]

Jefferson's words recall those in article 7 of the Articles of Association. Three and a half decades earlier, the delegates of the First Continental Congress had promised that those who could provide sheep to their poorer neighbors would do so on moderate terms.[47] Jefferson's use of similar language demonstrates that the concept of public virtue still held a warm place in the heart of this revolutionary.

After Jefferson expressed his outrage, he suggested a solution. Most of the remainder of the letter describes an elaborate scheme for selective breeding of Merinos, a proposal Jefferson advocated so that he and his friend, as virtuous Virginia gentlemen, could see that their state was well stocked with Merinos, provided free or at a reasonable cost. In this way, Jefferson hoped, he and Madison could undercut those who practiced what he called "scandalous extortions."[48]

A good deal of insight into Jefferson's political philosophy can be gleaned by noting what he did *not* write in the letter to Madison. There is no mention of any type of legislation—either by the United States or by the Commonwealth of Virginia—to regulate the sale of Merinos to curtail the selling practices Jefferson so despised. To Jefferson the way to end inflated Merino prices was not through any government action but rather through the private work of noble, civic-minded men. That, to Jefferson, counted as patriotism. Here, as elsewhere, his advocacy for limited government can be seen in how he thought about America's need for more sheep.

One thing Jefferson did see as a legitimate function of government was making war, and soon after his 1810 letter to Madison this would come. The Embargo Act of 1807, which Congressman Nicholson had addressed by way of his appraisal of domestic and foreign wool supplies, did not put an end to the festering hostilities between the United States and Great Britain. Five years later the two countries went to battle. Modern historians have generally been critical of President Madison's advocacy for taking up arms and Congress's subsequent decision to declare war.[49] But nearly a year and a half into the conflict, Thomas Jefferson wrote a letter to the Marquis de Lafayette defending America's actions—and once again the former president employed sheep to make his point. Jefferson boasted, "[Since the war's beginning,] the increase of our sheep is greatly attended to." His treasured Merinos in particular were "wonderfully extended, & improved in size," enabling the United States to produce "fine cloths . . . equal to the best English." Americans, he wrote proudly, were busy carding and spinning to make their own garments. "Were peace restored tomorrow," he asserted to his French friend, "we should not return to the importation from England of either coarse or middling fabrics of any material, nor even of the finer woolen cloths." With a flourish, he concluded, "Putting honor & right out of the question therefore, this revolution in our domestic economy was well worth a war."[50] Jefferson's nation, born in a war won in spite of the enemy's ovine advantage, now faced the same adversary—and the author of the Declaration of Independence was pleased to assure Lafayette that America presently stood on an equal footing with the British where quality and quantity of sheep was concerned. As will be seen in the next chapter, however, in praising Americans for spinning their own wool, Jefferson was expressing pride in a practice already in swift decline.

As noted, Alexander Hamilton wanted the federal government to award prizes for those raising superb sheep, a practice that in our time we associate county or state fairs. To hear Elkanah Watson tell it, such fairs were entirely his idea, although he acknowledged the help of a pair of sheep. A tireless promoter with multiple connections in business and in agriculture, Watson wanted to get in on the Merino craze. In the fall of 1807, he purchased two of the breed from Robert Livingston. Since Merinos were still a novelty,

Watson decided to make an impromptu exhibit of them. He tied his two prized sheep to an elm tree in the public square of Pittsfield, Massachusetts, and sent word out to the local farmers to come and have a look. That they did, some bringing their wives and families along.[51]

Seeing the farmers' fascination, Watson reported that he had a brainstorm. "I reasoned thus," he wrote, "If two animals are capable of exciting so much attention, what would be the effect of a display on a larger scale with larger animals?"[52] He shared this thought with some of the farmers assembled, and Watson later boasted that they agreed it was a fine idea and that, emboldened, he had organized such a display in 1810. It was a success, leading to the staging of an even larger agricultural exposition the following year. The day of the 1811 event, Watson recalled almost a decade later, "the village of Pittsfield was literally crowded with people, at an early hour, by estimation three or four thousand. Domestic animals were also seen coming from every quarter."[53]

Watson's tale of the sheep tied to the elm tree might be apocryphal. A copiously researched dissertation on the origins of American agricultural fairs discovered that no independent corroboration of the story has ever been found.[54] The same source charges that Watson had a decided tendency to exaggerate his own efforts toward establishing these fairs and of minimizing the contributions of the others involved.[55] But the event held in Pittsfield by the Berkshire Agricultural Society—of which Watson was one founder, if not the most important one—does have a compelling claim to be the first instance of what we would today call a county fair. While there had been earlier market fairs and agricultural exhibitions on both sides of the Atlantic, the 1811 Pittsfield Fair for the first time combined the two themes of education and entertainment. The displays of livestock, produce, and machinery served to disseminate knowledge of best practices in farming. At the same time, the fair featured speeches, parades, singing, and dancing, all serving to make the gathering a joyous festival.[56]

And there was a decided ovine connection to all this. Sheep were exhibited at Pittsfield, of course, especially prime Merinos.[57] More significantly, the establishment of the Berkshire Agricultural Society, hosts and promoters of the fair, was a natural outcome of the increasing importance of Merinos as the source of the raw material needed by the woolen mills beginning to arise in the area. After all, the founders of the society included men involved in both agriculture and manufacturing.[58] Sheep thus had an important role in the development of the rural fairs still held in America today—even if it can't be proven that Elkanah Watson ever tied two Merinos to an elm tree.

~~~

We have seen that influential men in the early years of the United States valued sheep, and that they expressed their feelings about the worth of these animals by sharing correspondence on their husbandry, by importing improved breeds, and, in Robert Livingston's case, even by publishing a book about them. Alexander Hamilton and Thomas Jefferson, who saw eye to eye on little, agreed that the young nation their words and actions had spawned needed to increase and improve its flocks, even though they disagreed on whether the federal government should be involved. It would be easy to conclude that not a discouraging word was heard at the time about ovines. There was, however, a very prominent contrarian who despised sheep and considered foolish the dreams of making the United States self-sufficient in its wool needs. He shared these thoughts in a best-selling and respected book.

John Taylor of Caroline—the geographical suffix is used to distinguish him from others bearing that common name—was a statesman, a political theorist, and a farmer. In all those things he reminds us of Jefferson and Madison. Unlike them, however, Taylor did not hold public office for as long, nor was he ever elected president, thus he is far less remembered today than his more famous Virginia contemporaries.[59] In his own time, however, Taylor was quite well known; one account declares that following the Revolution he "emerged rapidly as one of the principal citizens of the Old Dominion."[60]

Taylor authored several books, but when he is discussed today it is usually for his most famous one. In 1810 he began to write a series of essays for a local newspaper. These were gathered and in 1814 first published in book form, as *Arator*.[61] One biography of Taylor calls *Arator* "a handbook of both modern agriculture and democratic government."[62] These two things are not typically connected in our minds today. As a recent study put it: "[The] combination of political essays with discourses on manure, osage-orange fences, and pig-rearing seems odd to the modern reader."[63] But to Taylor, farming and politics were thoroughly intertwined. Like many prominent Americans of his time—especially Virginians—it was an article of faith to Taylor that the well-being of a republic depended on the virtue of its citizens, that agricultural pursuits especially fostered that essential virtue, and that, accordingly, an agrarian economy was necessary to preserve the United States as conceived by its founders.[64]

Arator was widely read, and its practical advice on farming was enormously influential.[65] It is doubtful, however, that the book's section on sheep had a significant impact. If it did, America would have given up raising sheep entirely. John Taylor absolutely loathed the beasts.

Taylor commenced his sheep essay by establishing his credentials. He reported that for sixteen years he had maintained a flock of one hundred to four hundred head, attended by a single shepherd. He concluded: "They require and consume far more food, in proportion to their size, than any other stock, . . . they are more liable to diseases and death, and . . . they cannot be made a profitable object throughout the whole extent of the . . . United States, but by banishing tillage from vast tracts of country."[66] Damning enough, but Taylor was just getting warmed up, leaping into hyperbole as he discussed sheep as veritable annihilators of civilization: "[If their] voraciousness is not gratified, the animal perishes or dwindles; if it is, he depopulates the country he inhabits. The sheep of Spain have probably kept out of existence, or sent out of it, more people than the wild beasts of the earth have destroyed from the creation; and those of England may have caused a greater depopulation than all her extravagant wars."[67]

In spite of the overheated rhetoric, Taylor's disapproval of sheep culture seemed at least in part based on sound economics and genuine concerns about land use. He conceded that wool was essential for keeping humans clothed. But, he argued, this need did not mean that Americans had to worry about having substantial domestic flocks. Taylor observed, obviously correctly, that the capacity of the earth to produce material for clothing is limited. Why then should Americans fuss with those pesky sheep when so much cotton can be grown here? Why not simply trade with England as better policy? "Under our warm and dry climate and in our sandy soil [we] can raise cotton," Taylor noted, "[while England, thanks to] her moisture and verdure, can raise wool cheaper than the United States."[68]

Taylor publicly expressed his objections to sheep when the *Arator* essays appeared, but privately he had shared his disdain much earlier. This is seen in a letter he sent to Thomas Jefferson in 1795, a reply to one he had received three months earlier.[69] In response to Jefferson's musing that he wished to divest his plantations from cattle in favor of sheep, Taylor respectfully suggested that Jefferson was making a big mistake. Taylor elaborated his thoughts in a two-column comparative chart, with "Sheep" the heading in one column and "Cattle" the heading in the other. In every point he deemed cattle superior. Sheep, Taylor wrote, were "liable to many fatal distempers," while cattle were "the most healthy domestic animal." In a related point, he declared that cattle were easy to fatten, sheep hard to fatten. Even the fleeces of sheep, providing that all-important commodity, wool, did not make them the equal of cattle, Taylor insisted, because one cow produces more meat and tallow than four sheep, giving bovines an economic edge.[70]

In spite of these reservations, Taylor wrote, he too was planning to try sheep on his plantation "as a means of turning turnips into manure." He did not say why he couldn't simply use cattle to turn turnips into manure, which is especially puzzling since he had just finished telling Jefferson that cattle manure was better. Amusingly, by the time Taylor wrote *Arator*, he disliked the vegetable he was feeding his sheep as much as the sheep themselves. Turnips, he concluded, were a worthless crop. "They are a food so little nutritious, that some animals die confined to them, none fatten without an additional food. . . . They are great exhausters of land, perhaps the greatest."[71] We might say that Taylor saw the turnip as the sheep's botanical equivalent.

When we note that his letter to Jefferson was written in 1795, it is difficult to avoid the impression that Taylor was being a bit disingenuous with his *Arator* readers. He had, after all, asserted in his 1814 book that he had maintained a flock of sheep over the preceding sixteen years, and he suggests that his negative impression about the animals was a result of his husbandry efforts over that period. But Taylor's letter to Jefferson, written nearly twenty years earlier, makes it clear that he had a dismissive view of sheep well before he had allegedly first undertaken their care. It would have been hard indeed for Taylor's sheep to overcome their shepherd's pessimistic preconceptions.

For decades after *Arator* was published, those advocating the establishment of vast flocks in the South criticized Taylor, suggesting that his sheep essay had led to a vast prejudice in the region against ovines, which in turn kept the population of southern sheep low.[72] But these complaints seem greatly overstated for several reasons. In the first place, as we have seen, Taylor's diatribe was, in its time, largely a voice in the wilderness matched against a thundering opposing chorus that exalted sheep. In Taylor's native Virginia, more powerful gentry than he were among America's best-known sheep boosters; this includes Washington, Jefferson, and Madison. Furthermore, even those who eagerly read *Arator* and tried to follow Taylor's instructions on other agricultural matters sometimes ignored his advice and tried sheep themselves.[73] As for Taylor's argument that sheep in the South did not make sense because the land was better suited for cotton, this became de facto policy anyway; it is unimaginable that either positive or negative opinions about sheep by any authority would have in the slightest way mitigated King Cotton's predominance.[74] Finally, as we will see in chapter 6, as the nineteenth century wore on, sheep numbers declined not only in the South but also in northern ovine strongholds such as New York and Pennsylvania, as America's livestock business moved ever farther west with each passing decade. That too was a development that was largely inevitable, regardless of what Taylor or anyone else on the eastern seaboard said about sheep. We will also see in chapter 6 that by the 1880s this westward movement of the flocks coincided with a popular revulsion against sheep that echoed Taylor's thoughts from seventy years earlier.

Given the concern Americans expressed between the 1770s and the War of 1812 about their dependency on sheep from Britain and Spain, it is striking that beginning in the 1820s the United States relied heavily on imported wool, and this came from sheep living in foreign lands far from western Europe. As we will see in the next chapter, a key reason for this turn of events was a practice that influenced much early American policy and development—slavery.

Embargos, Merinos, and the Flocks Grow

CHAPTER 3

"Utter Destruction to the Prosperity of That Section of the Union Whence I Come"

Sheep and the Rift between North and South

~~~~

> The state of feeling in the Union on the subject of the [wool] tariff... was the great question of difference between the two great sections of the country.
> —JOHN QUINCY ADAMS, 1832, speaking in the House of Representatives

When asked her age in July 1937, Cicely Cawthon figured she was about 78 years old. In spite of her advanced years, she took care of her mother-in-law, said to be 109. The two elderly women shared a small home in Toccoa, Georgia; it was there that Cawthon was interviewed about her enslaved childhood. "Master had sheep," she recalled. "In cold weather women wore woolen dresses and coats ... and woolen stockings, and the men wore woolen socks and britches too; everything they had was woven and made on the place."[1]

Cawthon's account was not unusual. Mary James, also interviewed in 1937, remembered that the farm where she had been enslaved had about 150 sheep for wool production. "The old women woved clothes," she reported. "We had woolen clothes in the winter and cotton clothes in the summer."[2] Nicey Pugh, born into slavery in Alabama in 1852, recollected washing, carding, spinning, and weaving wool. She stressed that the process encompassed the efforts of all the African Americans, from the men shearing the sheep to the children charged with picking the burrs out of each fleece. Pugh also told her interviewer that the enslaved people dyed their cloths; at her farm, cherry bark, dogwood, and gallberry were sources of these dyes.[3] Cawthon, James, and Pugh were all interviewed as part of the Federal Writers' Project (FWP), a Works Progress Administration (WPA) program conducted during the Great Depression. The FWP employed out-of-work writers, some of whom were sent to gather life stories from those who had been born into slavery.[4]

Plantation owners in the South also spoke of maintaining sheep and of processing wool into cloth on their own premises. In an 1809 letter to Thomas Jefferson, Georgia statesman John Milledge wrote that the typical planter in his state made his own cloth, "for his white family and Negroes." Earlier in the note, Milledge expressed his gratitude to Jefferson for providing a sheep, again remarking about his enslaved people: "I thank you for the Ice land ram, the wool from the breed of that animal, will answer for clothing our Negroes."[5]

For his part, Jefferson in 1812 wrote to John Adams touting the importance of his sheep as a source of raw material on his plantations. "[A Virginia operation like mine]," Jefferson asserted, "is a manufactory within itself, and is very generally able to make within itself all the stouter and middling stuffs for its own clothing & household use."[6] This philosophy was in keeping with Jefferson's hope, expressed a quarter of a century earlier in *Notes on the State of Virginia*, that America always remain a farming nation without need for manufacturing. "While we have land to labor," he wrote, "let our work-shops remain in Europe."[7] Sheep, by this thinking, were intended for domestic use, their wool homespun, not processed in a factory.

Nor was it only whites with large plantations and many enslaved people who satisfied their wool needs in-house. Historian Steven Hahn's thorough study of Georgia Upcountry yeoman farmers with few or no enslaved people showed that in the 1850s many of them raised sheep. Emphasizing that this was to provide wool for domestic use, Hahn reported that in certain counties over 70 percent of these families had their own spinning wheels and looms.[8]

Reading slave narratives, letters from white plantation owners, and statistics about small backcountry farmers, one could easily gain the impression that the antebellum South was self-contained in its wool needs, with generously large flocks of sheep providing all the fleece necessary to be made into winter clothing and blankets. In fact, this was not the case; the South had nowhere near the number of sheep required to meet its own demand. In particular, it was not the norm for the wool garments worn by enslaved people to come from sheep raised on their own plantation. This chapter will explain why and accordingly will lead us to examine how sheep were a factor in the conflict between North and South that eventually turned to bloodshed.

Silas Wright obviously had pride in the state he represented when he told the other members of the House of Representatives in March 1828 that recently New York had counted its sheep. Wright was serving his only term in the House, after which he became a U.S. senator, then the governor of New York. In 1844 he was offered the Democratic nomination for vice president, which he declined.[9]

Congressman Wright reported that New York's 1825 tally counted 3,496,539 sheep within its borders—he gave that exact number rather than rounding it off. He then pondered how many sheep there might be in the rest of the Northeast, which he defined as New Jersey, Pennsylvania, Maryland, Delaware, Ohio, and the New England states. Wright arrived at a figure of 10,313,139 sheep; added to the number in New York's flocks would give a total of slightly more than 13.8 million ovines in the region. Since the other states had not actually conducted a sheep head count, Wright had to extrapolate, basing his estimate on the assumption that the ratio of sheep to people in New York was similar to what would be found elsewhere. He admitted there was no firm reason to think this would be so, but he hoped by considering several states collectively the disparate ratios would even out.[10] He added an apology about his sum: "I am free to confess that this is vague and uncertain, but I have searched for better in vain."[11] Clearly for Wright counting sheep was a serious matter, although he probably overestimated America's ovine population—a later analysis concluded that in 1825 the entire country probably only had about twelve million sheep.[12]

Given the importance of sheep to the young American republic, it is surprising how long it took before there was a systematic count of the nation's flocks. As part of the 1810 U.S. Census, the Philadelphia economist Tench Coxe apparently made the first serious attempt to do this—but the result fell far short of what he no doubt had hoped for. Only five states, plus the Michigan Territory, submitted the returns Coxe desired. For what it was worth, those states and a territory collectively reported nearly 1.6 million sheep.[13] But even these meager returns were sufficient to cause Coxe to boast, "It is probable, that no country has ever effected so great a change in the value and extent of its stock of sheep, as the United States, within a very few years."[14]

Twenty-six years later there was another, far more thorough effort at a multistate sheep count, but rather than being a part of the government-mandated census, as was Coxe's 1810 work, this was a private endeavor. In 1836 C. Benton and Samuel F. Barry strove to precisely enumerate all the sheep in New England, New York, Pennsylvania, and Ohio; plus they estimated the size of the flocks in Virginia, Maryland, New Jersey, Delaware, and Kentucky.[15] They acknowledged that this was a formidable task. "Aware of the difficulty of obtaining correct information necessary to such a work," they wrote, "[we] have spared no pains to obtain that which may be relied upon."[16] Indeed, not content to simply present a state-by-state count, they published tallies of the number of ovines in each county and even for each township within a particular county. With a thoroughness that modern compilers armed with spreadsheets would admire, Benton and Barry reported that in Litchfield County, Connecticut, there were 2,914 sheep in Cornwall; 7,664 in New Milford; 8,999 in Salisbury; and so forth—for a countywide total of 72,832 sheep. Furthermore, Litchfield's wool, they wrote, was said to be "the finest in the State"—the fleeces were "well washed and put up in good order."[17] Tallying all the townships and counties, Benton and Barry determined that Connecticut's total flock consisted of 255,169 animals. Demonstrating that this was an effort to show the extent of an economy, they added to this count a note that the state had appraised its sheep flock at $378,482.[18] Adding their figures for all the states they canvassed, Benton and Barry counted nearly 12.9 million sheep.[19] But this was only fourteen of the states; they did not investigate how many sheep were in any of the states or territories south of Kentucky and Virginia or west of Ohio. This would happen a few years later in 1840, when, for the first time, there was a thorough effort to count all the nation's sheep.

Far more important than the question of how many sheep America had was the question of how many sheep America *needed*. In 1791 Connecticut legislator William Hillhouse calculated that it would take one million sheep just to provide clothing for the people of his state.[20] Connecticut had nearly 240,000 people at the time, so Hillhouse was advocating a ratio of more than four sheep for every person.[21] Thanks to sheep improvement, discussed in the preceding chapter, by 1845 Kentucky judge Adam Beatty computed a far more favorable ratio. He reasoned that the United States needed only about two and a half sheep for every human. Beatty figured that a typical sheep would produce two and a half pounds of wool yearly and that the

TABLE 3.1. Enslaved people and sheep in Cotton Belt states in 1840

State	Number of enslaved people	Number of sheep	Ratio of enslaved people to sheep
Alabama	590,756	163,243	3.6:1
Georgia	691,392	267,107	2.6:1
Louisiana	352,411	98,072	3.6:1
Mississippi	375,651	128,367	2.9:1
South Carolina	594,398	232,981	2.6:1
Totals	2,604,608	889,770	2.9:1

U.S. Census, 1840

average American had an annual consumption of about six pounds of wool, including clothing, blankets, and carpeting.[22]

Beatty did not mention the relationship between wool consumption and slavery. The New York statesman and agriculturist Henry Stevens Randall did, however, address this in his 1848 book *Sheep Husbandry in the South*. After mentioning Beatty's numbers, Randall noted that the average enslaved person in the South consumed half or less wool than "an ordinary Northern farmer or laborer, in comfortable circumstances."[23] Let us assume then, that if Beatty's average American of the 1840s consumed the wool of two and half sheep, that an enslaved person would be required to make do with only forty percent of this and received wool from only one sheep. How does this affect the impression we may have been given from the narratives at the start of this chapter, that, when it came to wool, southern plantations were self-contained, with the planter's own flock providing all that was necessary?

Very quickly, we see that this assumption is erroneous (see table 3.1). In 1840 there were slightly more than 690,000 enslaved people in Georgia. If each enslaved person was provided with the wool of one locally owned sheep, Georgia would be expected to also have a minimum of 690,000 sheep—and that number would be on the low end, as it assumes that every bit of wool clipped in the state went to enslaved people. In fact, the state only had around 267,000 sheep, a ratio of nearly one sheep for 2.6 enslaved people. In adjacent Alabama, the ratio was one sheep for 3.6 enslaved people. Table 3.1 shows the numbers for all five Cotton Belt states; regionally there were very nearly three times more enslaved people than sheep. Clearly there were not enough southern ovines to provide even a stingy allowance of wool used for the winter clothes that Cicely Cawthon, Mary James, and Nicey Pugh recollected. But if the sheep bearing the wool worn by enslaved people weren't in the South, where did they live?

⁂

At the beginning of *A Statistical View of the Number of Sheep*, Benton and Barry's book bore a dedication. "To the Hon. Henry Clay," it read, "the early and steadfast friend of the manufacturing and agricultural interests of this country."[24] Thus, the authors were showing their admiration for one of America's most prominent statesmen of the era—one who strongly supported tariffs on wool. Clay advocated duties on wool and other goods to support his "American system" of roads, canals, and other internal improvements.[25]

Clay was not alone in supporting wool tariffs. Which brings us back to Congressmen Silas Wright. When Wright strove to impress his fellow members of the House of Representatives with his sheep-counting oration, this was not a random musing. He brought up the sheep as part of the debate over one of the most contentious pieces of legislation in American history: the Tariff Bill of 1828, often referred to as the "Tariff of Abominations."[26] Wright was the chief author of the bill, and wool was at the forefront of the debate.[27]

Because the duties on iron, steel and lead were the first thing listed in the 1828 act, one could get the impression that metals were the foremost concern of the men in the twentieth U.S. Congress. This notion is likely to be buttressed in our minds as we think of all the things made of steel in our time—skyscrapers, automobiles, airplanes, computers, et cetera. But none of those things existed in 1828; even railroads were in their infancy.[28] Section 1 of the tariff is actually a reminder of the extent to which America then was still primarily an agricultural nation. The text included a roster of some of the metal goods affected by the law, including "sickels [*sic*] or reaping hooks, scythes, spades, shovels . . . [and] bridle bits of all descriptions."[29] Sheep shears were not specifically listed, but given their importance it would not have been surprising if they had been.

Nevertheless, in spite of their initial listing in the statute, it was not iron and steel that commanded the most debate. Instead, sheep and their wool weighed most heavily on the minds and the tongues of the members of the House of Representatives as they orated and argued. So much of the discussion focused on this that late in the debate an exasperated Missouri congressman complained that "the whole time had been consumed upon the single subject of wool and woolen fabrics . . . [while] the vast variety of other articles comprehended in the bill had been passed over in silence and oblivion."[30]

The congressional fixation on sheep had been evident the previous year, when the nineteenth Congress debated an ultimately unsuccessful tariff proposal.[31] Dutee Pearce of Rhode Island, presaging the 1828 remarks of Wright, estimated that there were fifteen million sheep in the United States. As shown, this was probably an overestimation, but the point of Pearce's lament was that America's sheep population was too small; he argued that the country was capable of sustaining a total flock of at least two hundred million. He figured that this number was easily attainable, arguing that America could double its flock in three years and quadruple it in six. Reflecting on this untapped potential disturbed him; he bemoaned that only about half the woolen goods manufactured in the United States were from American sheep. If it were not for the then-current economic downturn, Pearce declared, he would vote to ban all imports of foreign wool.[32]

Listing the particularly sheep-friendly states, Pearce mentioned his own Rhode Island plus Ohio, Vermont, Pennsylvania, and New York—all northern states. A few days later, William Archer of Virginia challenged Pearce's assertions. According to Archer, sheep required "short vegetation of rapid reproduction, of which the insular climate of Britain rendered large proportions of her soil prolific." These are conditions, he told his colleagues, not found in the United States. "No part of our country could be considered, perhaps, as in a very high degree favorable to the raising of sheep," Archer advised.[33]

Thomas Sill of Pennsylvania was flabbergasted; he said he had never before heard such an argument. In fact, he assured his Virginia colleague, probably "no country on earth was better adapted to the culture of wool, than the United States."[34] If Congressman Sill had read *Arator*, he would have heard the argument before. As noted in chapter 2, John Taylor, a Virginian like William Archer, had taken the identical position that America was unfit for sheep, but he argued that this deficiency was irrelevant, since the South had so much land unequaled for growing cotton. We see then the seeds for the emergence of a southern objection to sheep—why tie up good cotton land for sheep when flocks could be raised almost anywhere? In the North, climatically unsuited for cotton, no such prejudice against ovines arose. This difference between the regions had been emphasized earlier in the session when James Hamilton Jr. of South Carolina snarled that the chairman of the Committee on Manufactures—Rollin Mallary of Vermont—supported a high tariff as a result of "counting his ten millions of sheep that [might] have browsed on the mountains of the North" ("and ... calculating the quantity of wool which they carr[ied] on their backs").[35]

Considering such comments by southern statesmen, it is not surprising that the expansion of American flocks in the antebellum years took place mostly north of the Cotton Belt. And as we will now see, it was sheep that were not raised *anywhere* in the United States that were of particular concern to southern enslavers and a key to understanding their opposition to tariffs.

The talk in 1827 was just a dress rehearsal for the next year, when the Tariff of Abominations was passed. In 1828 the chatter went beyond grand estimates of America's sheep population. Several congressmen focused on the quality of wool, especially that of the lowest standard. New York representative James Strong praised sheep as "a species of capital." "Every farmer," he said, "can and should put on his farm as many prime sheep as he can conveniently keep." Strong promised, "In this way, every little flock will become a source of revenue to the owner, and all of them together of immense wealth to the nation."[36] His specification that the farmers should keep "prime" sheep as opposed to substandard ones is significant, because in the same speech Strong pointed out that wool was classified as coarse, middling, or fine, based on their quality, with coarse being the poorest material.[37] He then mentioned the source of this bottom-grade fiber: "This wool was principally from Smyrna and Buenos Aires and is a kind produced on an animal much inferior to the common sheep of this country; it is not grown in the United States and does not, in fact, come, in any way, in competition with any description of wool in the fleece shorn from the flocks of our farmers."[38]

With the place names he used, Strong was simply referring to the seaports they sailed from: the Turkish city of Smyrna (now called İzmir) and the capital city of Argentina. Most of the sheep providing this cheap wool lived on the steppes of Turkey or the South American pampas; sheared fleeces were then transported to the coastal cities of those lands for shipment. Strong's assertion that U.S. sheep did not produce this inferior fiber matched testimony heard earlier in the session by the House Committee on Manufactures. Joshua Clapp, a leading importer of wool, was asked by the committee what places he purchased from. The largest proportion of his imports, he responded, were very coarse wools that came from South America and Smyrna (see fig. 3.1). Asked next if those coarse wools came in competition with any of the domestic product, Clapp's answer was concise: "They do not."[39]

In addition to "coarse wool," "Smyrna wool," and "Buenos Aires wool," this low-grade material went by another name, one descriptive of its usual wearers. Clapp told the House committee that a factory in Canton, Massachusetts, used the imported coarse wool exclusively. "The cloth which they make is called and known as Negro cloths," he remarked.[40] Similarly, E. I. du Pont told the committee that his factory near Wilmington, Delaware, "used some Smyrna and some South American wool for Negro cloths, this wool being the coarsest kind."[41]

The answer to the earlier question where wool for enslaved people came from is now apparent: cheap wool was imported from Turkey and the Southern Cone of South America, specifically to be manufactured into cloth used to clothe America's enslaved people. This would be a significant factor in tariff policy and explains why southern planters opposed high duties on coarse wool, a commodity necessary to their operations and unavailable domestically.[42]

The Treasury Department had been reporting on wool imports for only a few years prior to Clapp's and du Pont's testimony. For the fiscal year ending in September 1822, the Treasury disclosed that 590,000 pounds of foreign wool classified as "coarse" had been shipped to the United States, with Spanish South America and the eastern Mediterranean the leading suppliers.[43] The onslaught of cheap wool imports became a tidal wave as the antebellum years went on. For the fiscal year 1846, the United States imported over 16.4 million pounds of bottom-grade wool, almost twenty-eight times more of this fiber than reached American ports in the early 1820s. The two leading sources of this coarse wool again were Turkey, contributing over 5.7 million pounds, and Argentina, providing almost 4.3 million pounds.[44] Looked at in another, rather ironic way, 16.4 million pounds is 8,200 tons. The Statue of Liberty tips the scale at only 225 tons.[45] Thus the wool imported in just one year, largely for the use of enslaved people, weighed over thirty-six times more than America's most prominent monument to freedom. High levels of such imports persisted even when slavery neared its end. In 1865 nearly 17.3 million pounds of the lowest grade of wool were brought into America's ports, suggesting that the woolen manufacturers believed

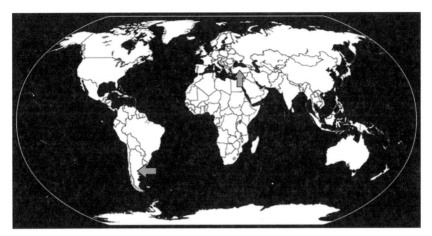

FIG. 3.1. Most wool used to manufacture "Negro cloths" between the 1820s and the U.S. Civil War came from sheep living in the Southern Cone of South America or in Turkey.

that a market for the very cheapest cloths would persist even in the postwar economy.[46]

In contrast, imports of higher-quality fleeces were minimal before the Civil War. When 16.4 million pounds of coarse wool came to America in 1846, only about 130,000 pounds of more-expensive grades arrived—in other words, about 99.8 percent of the imports were low-grade wool, a mainstay for enslaved people's use.[47]

In the 1827 and 1828 debates, Congressmen Wright and Pearce glowingly spoke of the vast flocks of sheep in the Northeast. But the coarse wool used for enslaved people did not come from these animals or from any other U.S. flocks. With the market for "Negro cloths" in the South reliable and expanding, why didn't sheep farmers in New York, Pennsylvania, and other northern states supply this demand?

The economics of domestication was a key factor. *The Oxford Dictionary of Zoology* defines "domestication" as "the selective breeding of species by humans in order to accommodate human needs."[48] In the case of the sheep, that selective breeding has sometimes been for mutton, but more often it has been for wool quality. Regardless whether a sheep's wool is luxuriantly soft or irritatingly coarse, it costs the same to purchase fodder or to build a barn. American shepherds thus were not interested in keeping sheep with coarse wool when for the same expense and greater profit they could raise

Utter Destruction to the Prosperity

animals with higher-quality fleeces. This, several congressmen explained in the 1828 debate, was the reason wool for enslaved people came from abroad. Typical were the comments of Isaac Bates of Massachusetts, who said of sheep with coarse wool, "What [American] farmer would think of stocking his farm with them? He may keep the fine wool sheep as cheap, and they will produce him double and treble the money."[49]

In *Sheep Husbandry in the South*, Henry Randall criticized the South for its dependence on foreign sheep and distant factories "for the apparel and bedding of slaves."[50] Randall considered this bad economics; he concluded that the South would be better off growing and manufacturing its own wool. Quite implausibly, however, he further argued that rather than settling for sheep with coarse wool, such as was being imported, southern agriculturists should instead concentrate on breeding sheep with a higher grade of wool. "Even . . . for the uses of the plantation—for slave-cloths, etc.," Randall asserted, "*Fine wool is worth more per pound than coarse for actual wear or use!*"(emphasis his)[51] It hardly seems probable, however, that a plantation owner, careful in his selective breeding to cultivate flocks with high-quality fleeces, would be eager to use such wool for his enslaved people. More likely such a planter would do a quick calculation and conclude that if he had sheep with fine wool, he would be better off economically to sell it at a profit and continue to buy cheap cloths for his enslaved people. Furthermore, even if there was no economic consideration, there would likely be a sociological one: the planter probably would not wish to see his enslaved laborers wearing wool as good as that he himself sported. This observation goes the other way as well: white Americans of the antebellum era looked down on coarse wool because of its association with slavery. As Silas Wright said during the 1828 tariff debate, "The Negro cloths cannot and will not be worn by the Northern laborers. Neither their climate nor their feelings will permit it."[52]

Wright and other supporters of the proposed tariff heard howls of disagreement from southern representatives opposed to the tariff bill. North Carolina congressman Daniel Turner strenuously objected not only to the bill under consideration but also to the earlier acts already in force. Northern representatives supporting the industrialists should tread lightly on the matter of tariffs, Turner warned. "[If they rise too high] we shall be too poor to purchase their manufactured articles." He complained: "Many of us are [already] driven to the necessity of manufacturing our own Negros' clothing; those hands that were formerly more profitably employed in the field, are turned, by the operation of your laws, to the spindle and loom, to save the expenditure of that sum which was given by the planter for the clothing of his hands, and given freely, because he could afford it."⁵³

Less than two decades after Thomas Jefferson and John Milledge extolled the virtues of clothing made from scratch on their plantations, here was a representative from a slave state horrified at such a prospect. Why tie up valuable labor for such a thing, Turner argued, if you can buy cloth already produced at a factory? Turner did not specify in his lament whether those enslaved people he described were making clothing of wool or cotton or both. But the key point is his assertion that it was bad, unprofitable management to have enslaved people spinning or weaving at all when they should be planting, weeding, or harvesting.

Turner's speech was an unusual case of a southern politician, in a roundabout way, acknowledging his region's dependence on the northern factories—he even specified the reliance of slaveholders on manufactured cloth to clothe their enslaved people. This was not a frequently expressed sentiment; southern statesmen were far more likely to denounce the manufacturers than to admit that their existence benefited the slavocracy. Just a few days after Turner spoke, South Carolina representative William Martin took the floor and decried efforts "to levy a tax on the Southern people for the benefit of manufacturers in the middle and eastern states."⁵⁴ Four years later, when the House debated revisions to the 1828 act, another South Carolinian, William Drayton, presented a memorial from the legislature of his state, complaining that "the benefits of the tariff [were] confined to the manufacturing States, and that South Carolina . . . receive[d] no part of the compensation."⁵⁵ But as Congressman Turner perhaps inadvertently admitted, in a real sense the slave states did receive compensation.

For nearly thirty years following ratification of the Constitution, the United States had no duty on wool. But as noted in the last chapter, after the War of 1812 the British flooded the U.S. market with woolen cloth and clothing—not only depressing prices and causing sheep farmers to slaughter many of their sheep, but also harming efforts to develop a domestic industry.⁵⁶

Congress reacted in 1816 by enacting America's first tariff to include wool; it was a 25 percent ad valorem assessment on all manufactured woolen cloth. This was supposed to drop to a 20 percent tax after three years, but the scheduled reduction was indefinitely postponed.[57] The new law specified that, with a few exceptions, the tariff was to apply to "woolen *manufactures* of all descriptions" (emphasis mine). Unprocessed wool was not mentioned, rendering it questionable whether there was a legislative intent to subject it to customs duties. But in addition to a list of items covered by the tariff, the act also contained a section enumerating things that could be imported duty-free—and raw wool was not among those commodities and goods listed.[58] Thus excluded from the free list, raw wool fell under the category of "all articles not free, and not subject to any other rate of duty."[59] These things were assessed a 15 percent tariff, which is what importers of raw wool paid.[60]

Why the oversight, if it was an oversight? Probably it was because when Congress passed the 1816 law, imports of unprocessed wool were insignificant. The Treasury Department reported that during the following year, 1817, less than twenty-three hundred pounds of raw wool came into the country.[61] As noted earlier, that would change in a few years, with imports of unprocessed wool—especially the cheapest kind—swiftly rising. This would have a profound effect on further tariff acts.

For a variety of reasons—including the continued stream of manufactured British woolens in spite of the 1816 tariff—the U.S. economy collapsed in 1819. Prices of American goods and land values fell by 50 to 75 percent. "Backward the dominos fell," historian Sean Wilentz put it, "from ruined speculator to merchant to farmer."[62] The so-called Panic of 1819 arrived when the American woolen manufacturing business was still in its infancy, and a number of factories went out of business. Those remaining earned little or no profit.[63] Unsurprisingly, those surviving establishments sought to obtain further security from imports, and the resulting protective movement deeply influenced American politics for years.[64]

Advocates for a new tariff eventually prevailed in 1824, enacting a law that in two stages raised the ad valorem duty on manufactured woolens to 33.3 percent (see table 3.2).[65] But there was another change from the 1816 legislation—the new law specifically mentioned raw wool and set rates for its importation. Unprocessed wool worth over ten cents a pound would pay an ad valorem tariff of 20 percent, while the rate would be only 15 percent for wool worth ten cents a pound or less.[66] That cheaper wool was precisely what was being imported from Turkey and Argentina to be manufactured into "Negro cloths." Note that the 15 percent duty in 1824 was the same as it was in 1816—but those raw wool imports were rapidly increasing to meet

TABLE 3.2. Tariffs on coarse, unmanufactured wool in the United States from the nation's founding through 1830

Before 1816: none
From 1816 to 1828: 15 percent ad valorem
Example: For 12,500 pounds of wool worth 8 cents a pound, the value is $1,000. The importer pays **$150**.
1828, the "Tariff of Abominations": 4 cents a pound, plus 40 percent ad valorem—rising to 50 percent in 1830
Example: For 12,500 pounds of wool worth 8 cents a pound, the value is $1,000. The importer pays $500 (.04 x 12,500 plus $400 (40 percent of 1,000) for a total duty of **$900**. This would rise in 1830 to **$1,000**.

the needs of expanding southern slavery. Northern factories purchasing this coarse wool passed the tariff cost on to their slaveholding customers. This was the burden Daniel Turner complained of when he insisted that it was causing planters to unprofitably work enslaved people at the "spindle and loom" instead of in the field.

But, in spite of Turner's plea for a reduction, the 1828 bill instead raised the rates. Importers of unmanufactured wool would from that point pay four cents a pound plus an ad valorem of 40 percent, with scheduled raises over the next two years bringing the ad valorem up to 50 percent.[67] Table 3.2 shows the effect of this increase; for the same amount of wool, a duty of $150 before the Tariff of Abominations rose to $900, with a further scheduled increase bringing the rate to $1,000. As the table shows, in effect this meant that an import of coarse wool worth one thousand dollars would be subject to a thousand-dollar tariff. Not only was this a massive increase in the tax, but gone was the 1824 discount for the coarse wool. The same duty would be paid on raw wool whether it came from the finest flocks of Saxony to adorn America's rich or from the derelict flocks of Argentina or Turkey to be worn by southern enslaved people.

Southern congressmen enraged by this turn of events had some allies in the north. In the 1828 debate, Pennsylvania representative James Stevenson pleaded that his colleagues make sure there was no undue burden on slave owners wishing to do the right thing and keep their enslaved laborers properly clothed. Stevenson asked: "[Will those sympathetic to the enslaved people] . . . forget that it may be an act of wide-spreading practical humanity to keep down the value of the clothing usually bestowed? Can they who are

Utter Destruction to the Prosperity

preaching philanthropy forget their own doctrines, and tempt the master to stint the man?"[68]

Stevenson's arguments fell on deaf ears in 1828, but over time a feeling arose among other northern statesmen that the new rate was much too high. Four years later, when Congress reconsidered the wool tariff, more delegates from free states supported a reduction or elimination of the duty on raw wool as a reasonable concession to the South. "That kind of goods peculiarly used at the South was one on which the duties might easily be remitted," John Quincy Adams declared in the 1832 debate.[69] His fellow Massachusetts representative Rufus Choate agreed, hoping that articles "exclusively of Southern consumption" would be made duty-free, including "Negro clothing."[70] One might hope that these New Englanders were at least partly driven by the humanitarian impulse James Stevenson advocated—especially Adams, widely praised today for his roles as opponent of the congressional gag rule on petitions for ending the District of Columbia slave trade and as legal counsel when the *Amistad* rebels sought their freedom.[71] A more cynical argument is that Adams and Choate represented a state where much of the raw foreign material was manufactured into cloth; they thus had a provincial interest since a reduction in the tariff on coarse wool would benefit their constituents as much as the enslavers.

What effect had the Tariff of Abominations actually had on raw wool imports? In 1829 imports dropped by about 28 percent from the previous year. The next year, however, with the tariff still in effect, the raw wool brought in increased by 4.2 percent. The steep drop in 1829 could partly be explained if manufacturers had been stockpiling wool in preparation for a possible tariff increase. This would also explain the modest uptick in imports in 1830. As the stockpile diminished, more wool had to be imported to meet the demand in spite of the increase in duties. These figures should be interpreted with caution, however. With wool, as with all commodities imported under the 1828 tariff, there were wide allegations of impropriety and fraud; obviously this would have compromised the accuracy of the Treasury Department's statistics.[72]

In 1832 a new tariff act was passed, keeping the rate for unmanufactured wool at four cents per pound and 40 percent ad valorem, with an important exception. Any raw wool worth eight cents or less per pound at the place of exportation would now be admitted duty-free.[73] The coarsest wool, used almost entirely to clothe enslaved people in the South, was no longer subject to a federal tax. As a result, imports skyrocketed; between 1835 and 1841

TABLE 3.3. United States imports of raw wool, 1828–1830

Year	Pounds imported	Change from previous year
1828	1,378,170	—
1829	992,540	Down 28%
1830	1,035,557	Up 4.2%

U.S. Census, 1840.

the amount of raw wool brought into the United States increased by 250 percent.[74]

While this seemed to be a major concession, it was not enough to satisfy some in the South, especially in South Carolina. Several months after passage of the 1832 tariff law, that state's legislature passed the Ordinance of Nullification, declaring that the 1828 and 1832 tariffs each were "null, void, and no law."[75] While the ordinance did not mention wool specifically, it accused Congress of "wholly exempting from taxation certain foreign commodities, such as are not produced or manufactured in the United States, to afford a pretext for imposing higher and excessive duties on articles similar to those intended to be protected."[76] Coarse wool met all those definitions; it was a foreign commodity not produced in the United States, and the 1832 tariff had exempted it from taxation. South Carolina's attempt at nullification thus can be read to imply that the removal of the duty on coarse wool was not a favor to the slavocracy but rather a bad-faith effort to increase other tariffs. The rift between the North and the South widened.

In recent years, scholarship on cotton has emphasized its centrality to the global aspect of antebellum slavery—a southern slave picked the cotton, which was then transported to textile factories and then made into consumer clothing.[77] With coarse wool, globalism was no less prominent, but here the enslaved people stood at the end of the chain rather than at the beginning. The fleeces shorn from sheep in the pampas of South America or the steppes of Turkey were transported to Buenos Aires or to Smyrna, respectively, then loaded onto ships taking them to eastern U.S. factories where the wool was processed into "Negro cloths" that were sold to southern planters, who then issued it to their enslaved workers (see fig. 3.2).

FIG. 3.2. Engraving of a textile factory in 1836, loom on the right producing finished cloth. From George S. White, *Memoir of Samuel Slater*, 2nd ed. (Philadelphia: Gales & Seaton, 1836).

We are now in a position to revisit the interviews with former slaves presented at the beginning of this chapter. If Cicely Cawthon was seventy-eight in 1937, she would have been born around 1859. In other words, hers were the recollections of an elderly woman whose earliest memories of life would have come during the Civil War, when slavery neared its extinction. This was obviously not a period of business as usual in the South; rather, it was a time of the Slave Power fighting for its survival. It is very likely that the world Cawthon, Mary James, and Nicey Pugh described, of the wool they wore coming directly from the sheep on their plantations, was a wartime expedient, not typical of what would have been the case just a few years earlier. Yes, some planters had sheep on their property; there is no reason to doubt James's comment that there was a flock of 150 where she lived. But those sheep were probably improved breeds, producing fleeces of a higher quality than the plantation owner would have wished to employ for slave clothing.[78] Instead, he would have used this prime wool for trade or sold it for profit. In the waning days of the Old South, a planter unable to acquire foreign goods may indeed have allowed or even instructed enslaved people to make their garments from his prized sheep's wool. But this would have been a policy born of wartime realities, not of peacetime wishes. Typically, antebellum slaves did not face the winter wearing garments made of wool from the sheep living a few hundred yards from their cabins. Instead, they wore clothes or slept under blankets derived from wool shorn off the backs of sheep living thousands of miles away.

And what of the words of Thomas Jefferson and John Milledge, so steadfast in their assurance that a plantation was self-contained in its wool needs, with the fleece from the owner's sheep carded, spun, and turned into clothing for, as Milledge put it, the white family and the Negros? While Cicely Cawthon's reflections were from late in slavery's existence, Jefferson and Milledge were describing an earlier America, just prior to the War of 1812—before the peak of slavery and, more important, before domestic factories became significant. In 1810 Treasury Secretary Albert Gallatin reported that in the United States almost all wool was "spun and wove in private families" ("there are yet but few establishments for the manufacture of woolen cloths").[79] Economic historian Chester Whitney Wright estimated that at this time at least twenty-four of every twenty-five yards of woolen cloth produced in America was made in the household.[80]

This was about to change. The handful of woolen mills Gallatin's 1810 report noted would soon have serious competition. E. I. du Pont was at work building his woolen factory in Delaware, the same factory that he referred to when testifying before the House Committee on Manufactures in 1828.[81] In the spring of 1812, du Pont received a letter from an eager potential customer. "I have understood you are concerned in a manufactory of cloth," the note began, "and will receive one's wool, have it spun, wove & dyed for an equivalent in the wool . . . will you be so kind as to inform me more particularly on this subject."[82] The writer of this letter was Thomas Jefferson—the same Jefferson who in the 1780s expressed his hope that "work-shops" would remain in Europe. In a few weeks, du Pont sent a response, assuring Jefferson that "our manufactory is entirely calculated for the use of fine Merino wool and we will with great pleasure manufacture the fleeces of your full blooded sheep, for your own account, if you desire it."[83] Pleased, Jefferson wrote back: "I am happy to know that we have established among us a manufacture from which we may expect to see the French processes in both weaving & dying fine cloths introduced among us."[84] This was a far cry indeed from Jefferson's earlier wish in *Notes on the State of Virginia* that no one in America would be "occupied at a work-bench, or twirling a distaff."[85] We should also note du Pont's remark to Jefferson that his factory was strictly concerned with fine Merino wool—he was not yet importing the coarse wool for Negro cloths that he would tell the congressional committee about sixteen years later. Ultimately, du Pont wound up not only mass-producing woolen cloth but also importing the raw material, another thing Jefferson had decried three decades earlier in *Notes*.[86]

While tariff debates inevitably revolve around the specific goods to be taxed and at what rate, a chief issue regularly raised after 1820 centered on the stated purpose for the duties. Were they simply for raising revenue to operate the government, or was their goal to protect American manufacturers from foreign competition? While the power of Congress to raise revenue was generally not questioned, its authority to support domestic manufacturing increasingly met objections from those who argued this policy was not only unsound but also unconstitutional. Among statesmen from the South, these protests were especially vociferous.[87] Historian Daniel Walker Howe argues that the difference between a tariff being considered "for revenue" or "for protection" was largely irrelevant, as it was largely in the eye of the beholder. "All tariffs have both elements," he explains, "and the congressmen who voted for it probably did not all have the same motive."[88]

No doubt this is true. But when it came to the specific matter of low-grade wool, it is hard to avoid the conclusion that the tariff laws of 1824 and 1828 had nothing to do with protection. Since virtually none of the coarse wool used to make "Negro cloths" came from American sheep, having any duty at all on cheap, raw wool wasn't protecting the economic interests of anyone in the North—neither the sheep farmers, who did not want to raise sheep bearing inferior fleeces, nor the woolen manufacturers, who had no domestic supply for low-grade wool and would have to seek it abroad with or without a tariff. On the other hand, the southern complaints about northern manufacturers were disingenuous; the slavocracy depended on those factories for finished woolen cloth to provide their African American labor. Congressman Turner's lament that high tariffs were driving planters to waste resources by having enslaved people spin and weave their own cloth—whether this was actually happening or not—was far closer to an expression of the South's interest than any diatribe against New England factories. In many ways, then, the North and the South were in these debates talking past each other.

Furthermore, since there was no protection involved where coarse wool was concerned, the tariff on it could be accurately described *only* as a revenue-generating measure. This too may have been galling to the South—essentially, customs on cheap wool were a tax on slavery itself. Perhaps irritation over this was one of the things on Georgia representative Wiley Thompson's mind when he exclaimed during the 1828 debate that the proposed tariff bill threatened "utter destruction to the prosperity of that section of the Union whence I come."[89]

In antebellum America, wool tariffs and slavery were inseparable. The South needed wool to clothe their enslaved people, and their own region did not have enough sheep to supply this demand. Nor were the enslavers willing to buy cloth manufactured of wool shorn from northern sheep, because the farmers there maintained flocks bearing wool of a higher quality than the planters were willing to pay for. Thus the South depended on cheap, coarse, imported wool, a commodity that was deeply affected by the 1820s tariff bills. In part, the widening rift between North and South was due to the influence of sheep on American economics. And when at last blood was shed, a South without wool to clothe its slaves would also be without enough wool to clothe its armies fighting to preserve slavery.

With secession and the outbreak of the Civil War, the South's dependence on foreign trade and northern manufactures for its wool led to a new problem. How would the South procure sufficient wool for military use? Accordingly, when the legislature of South Carolina, the first state to secede, assembled in November 1861, a message from Governor Francis W. Pickens was read in which he declared: "The State should . . . give encouragement to raising and manufacturing wool enough for our necessary wants."[90] The next month, both houses of the state legislature concurred in supporting a report from the Committee on Commerce and Manufactures "as relates to the manufacture of cannon, small arms, and gunpowder, and the raising and manufacture of wool within the State" (note the linking of wool with weaponry).[91] When armies clashed, sheep were munitions.

In 1860 the United States had nearly 22.5 million sheep.[92] Only about 5 million of these were in the states that seceded the following year, so the Union had a decided advantage when war broke out. But the North too strained under wartime demands. A complete outfit for a Union soldier

FIG. 3.3. A typical Confederate soldier's uniform consisting of a shell jacket and trousers made circa 1863 at the Wytheville, Virginia, clothing depot. Atlanta History Center collection, accession number 2005.200.513.

required approximately twenty-five pounds of raw wool, and given wear and casualties it was estimated that the annual consumption amounted to around sixty pounds per man in service. For the year 1864, the Union army's total wool consumption had reached about seventy-five million pounds. Northern factories ran round the clock and remained operational on Sundays to meet the increased need for woolen clothes and blankets.[93] Imported wool helped furnish America's unprecedented demand, but the crisis was also met through a renewed commitment by sheep farmers to increase their flocks. The Civil War, in effect, doubled the number of American sheep, vastly increasing the supply of wool. Chester Wright calculated that in 1865, the year the war ended, the United States had a total wool clip of 155 million pounds, nearly twice the estimated 80-million-pound clip of 1860.[94]

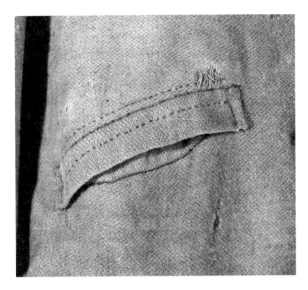

FIG. 3.4. This detail of the Wytheville uniform shows the cotton warp in the spot above the pocket where the wool was eaten by moth larvae.

The Confederacy, on the other hand, never was able to increase its wool supply as Governor Pickens had hoped. Following the Battle of Gettysburg, Samuel Weaver, supervisor and recorder of the exhumations, reported that one way of determining whether a body was Union or Confederate was to note not just the color of the uniform—gray or brown for the Rebels, blue for the Union—but also its material. Rebel clothing, he said, was made of cotton. Not so for the Union: "The clothing of our men is wool."[95] But Weaver oversimplified. As he noted, Union uniforms were typically all wool, but Confederate wear generally was a blended fabric. The South stretched its thinner supplies by manufacturing uniforms composed of "jean cloth," a fabric woven with a cotton warp and a woolen weft (see figs. 3.3 and 3.4).[96]

The end of the Civil War meant that the United States would no longer fight over slavery. But there would still be fights over sheep, the source of that all-important wool.

Utter Destruction to the Prosperity   55

CHAPTER 4

Shepherds Enslaved and Free
Sheep and Race

> Let sheep culture, upon ever so modest a scale, generally prevail among the colored people of the South, and [sheep-killing] dogs, and thieves, white or black, would quickly disappear under the vigilance of a self-constituted police, more effective than the law could provide.
> —JOHN L. HAYES, "Sheep Husbandry in the South"

The black residents of Caroline County, Maryland, were excited. Reverend Peter D. W. Schureman of the African Methodist Episcopal Church (AME) was coming! How they looked forward to hearing him speak. If his reputation was deserved, he was the greatest orator they might ever hear. Naturally, they made extensive preparations for the reverend's highly anticipated visit in the spring of 1826—or was it 1827? Writing about the big event over fifty years later, Alexander W. Wayman—who himself later became an AME bishop—could not remember which year the visit occurred. But whatever year it was, he figured it must have been in May, "for when he came my father was away from the house shearing his sheep. I was there with him."[1] Clearly shearing was a notable occasion for the Wayman family; this would explain why over five decades later Wayman's mind connected the time of the wool harvest with Reverend Schureman's visit. Also notable is the possessive pronoun used to describe the animals—Wayman's father was shearing "his" sheep, not those of a white owner. The Waymans were a free family. In 1820, the year prior to Alexander Wayman's birth, Caroline County had a population of 1,574 enslaved, but also 1,390 free people of color.[2]

The preceding chapter considered how wool tariffs caused sheep to be a part of the sectional rift developing in America during the antebellum years, when procurement of low-cost wool became a major focus of the slavocracy. This chapter more directly addresses sheep as an element of the relationship between white and black America, in both the antebellum and the post–Civil War years. Regrettably, firsthand accounts by African American shepherds are scarce; it is thus unavoidable that much covered here is gleaned from the perspectives of European Americans.

Agriculture-based slavery in the United States is usually associated with useful plants. There are numerous accounts of the work done by enslaved workers in fields of tobacco, rice, sugarcane, and especially cotton.[3] But enslaved people were also commonly tasked with caring for sheep and other livestock.

Sheep were first domesticated in the Middle East and brought to the New World from Europe. Given the proximity of Africa to their place of origin, it is not surprising that sheep arrived in Africa far earlier than their first transport to America—they were in Saharan regions by eighty-five hundred years ago. As they were carried south into tropical Africa, distinct breeds were developed that lacked wool and were raised strictly for their meat—a logical development in the torrid zones where there would be little use for wool clothing. Unlike the American Indians, then, many Africans in their first contacts with Europeans were already familiar with sheep.[4]

Transatlantic slave ships sometimes carried sheep as a food source, reflecting the concerns of the merchants operating the slave trade that their human chattel arrive in the New World in reasonable health (or, as they would think of it, in saleable condition). One eighteenth-century captain of such a vessel received specific instructions to keep sheep and goats on board to make "mutton broth" to feed enslaved people when they were ill.[5] Some of those who served the broth, and some of their descendants, would later take care of the living source of the mutton.

"Sir," Joseph Dougherty politely asked in a letter he sent to Thomas Jefferson in 1809, "it may be that you have some old Negro that is of little or no use to you. If such you should have, and would think it right to send one to me to take care of my sheep; I will agree to give you any reasonable compensation you would ask, for such a man I cannot find in this place."[6] Jefferson's term as president had ended just a few months before he received this communication from Dougherty, who had served as presidential coachman and head of stables while Jefferson held office.[7] After leaving Washington, Jefferson remained in touch with his former employee. Much of their correspondence concerned acquisition and husbandry of sheep. In the letter quoted here, the key point is Dougherty's request for a slave—but just one, and he could be elderly and of little other use—to handle sheep husbandry. This stood in contrast to the number of hands needed for plant-based agriculture.

The rapid expansion of slavery in the early nineteenth century was a direct result of the corresponding explosion in cotton production.[8] Horrifying firsthand accounts exist of the arduous labor imposed on those working the cotton fields, stressing the number of enslaved put to work, the difficulty of their task, and the severity of the punishment when one fell behind in the mandated workload—or was simply perceived as not laboring hard enough. Charles Ball reckoned that on his first day of lifting a hoe to weed a South Carolina cotton field that he was one of 168 enslaved forced to toil.[9] Similarly, Louis Hughes, writing of his experience as a slave harvesting cotton in Mississippi, reported that every able-bodied slave was expected to pick 250 pounds a day or more, and that "all those who did not come up to the required amount would get a whipping."[10]

Compare these details from survivors of cotton-based antebellum slavery, emphasizing the size of the workforce and the backbreaking nature of their toil, with Dougherty's request for a single "old Negro" to handle the shepherding. Dougherty's notion that sheep required a small outlay of labor was no outlier; others also asserted that a single shepherd—perhaps accompanied by a well-trained dog or two—was all the staffing needed to maintain hundreds of sheep.[11] In the antebellum South that shepherd would likely be African American, like the man that Dougherty sought. Writing to his farm manager in 1793, George Washington declared that he had thought of consolidating the over five hundred sheep at all five of his farms into one large flock, minus rams, and having the whole large group "placed under the care of a trusty negro . . . whose sole business it should be to look after, and fold them every night."[12]

The labor-saving differences of sheep compared to cotton are self-evident. The huge task of eliminating weedy competitors from cotton rows, described by Charles Ball, is not a concern in the raising of sheep. Harvesting also required fewer hands. Recall Alexander Wayman's description of his childhood; when his father sheared the sheep, he, as a boy, assisted. If only the two of them were present for the work, this is congruent with the recommendations of nineteenth-century sheep husbandry manuals. Typically these state that besides a shearer—or shearers, if there are a very large number of sheep—the only other person needed is a "catcher" to deliver the sheep to the shearer. This is a far cry from the hundreds of hands sent into antebellum fields to pick cotton.[13] The minimal labor required for wool compared to cotton growing meant that the number of African Americans working as shepherds would always be a tiny fraction of those planting, weeding, and harvesting on southern plantations.

Economically, not only was the amount of labor needed to maintain sheep less than that required to grow cotton, slavery also made it cheaper to keep ovines below the Mason-Dixon Line. In *Sheep Husbandry in the South*, published in 1848, the agricultural writer Henry S. Randall calculated that the gross cost of producing a pound of wool in New York was over twenty-seven cents, compared to only about eight cents in the South. This, Randall emphasized, was mostly a difference in labor—each shearer in the North charged a dollar a day, but slavery eliminated this cost in the South. Furthermore, for day-to-day sheep care, healthy adult enslaved people were not needed. The task "could be borne mainly, if not entirely, by superannuated or decrepit slaves, or even by children."[14] Again we see an emphasis on sheep care requiring few hands, hands that did not need to be robust.

Given the significance America placed on its flocks of sheep, it was inevitable that it would affect the view the nation's white people had of African Americans. This would become a matter of politics as well as of economics.

Historian David Brion Davis notes that equating enslaved people to livestock is an ancient practice—Sumerian tablets from the middle of the third millennium BCE correlated slaves with domestic animals in the manner in which they were priced, graded, and described.[15] The comparison was thus purely economic—both the enslaved and the stock were viewed in terms of their value to those possessing their services. Writing of the situation in antebellum America, the nineteenth-century Scottish journalist Robert Somers made this point by contrasting the South with his native Great Britain. He wrote, "It was the property in slaves that gave to the Southern plantations permanence of value . . . the substance that a British farmer possesses in his sheep and cattle."[16]

That connections between sheep and slavery could take on a political perspective as well as an economic one is shown by an odd exchange that occurred when the Second Continental Congress debated in the summer of 1776. After voting in favor of separation from Great Britain and concurrently adopting the Declaration of Independence as their official document justifying the break, the assembled delegates confronted another difficult task—finding agreement on how thirteen separate states could organize a collective government. Late in July, they debated establishing tax quotas for each state, including taxation to be incurred by slaveholders versus non-slaveholders. Hearing a northern delegate assert that slaves should be assessed more than other types of property was too much for Thomas Lynch Jr. of South Carolina. Why, he bellowed, should slaves be taxed more than land, sheep, cattle, and horses? Benjamin Franklin responded. "Slaves rather weaken than strengthen the State, and there is therefore some difference between them and Sheep," he insisted, adding, "Sheep will never make any insurrections."[17]

Two points stand out about Franklin's brief retort. The first is that while Lynch brought up three different species of livestock—sheep, cattle, and horses—Franklin focused on the sheep as the one that, in contrast to slavery, is a source of a republic's strength. The second is the humor of the remark: by conjuring up the absurd image of an insurrection of sheep, Franklin spiced the argument with his wit. It is easy to picture the other men assembled that day chuckling at the comment. Perhaps a few laughed out loud.

But forty-six years later Nicholas Herbemont wasn't laughing.

Today an American who enjoys a glass of wine with dinner may, for reasons of taste, patriotism, or both, select wine made from grapes grown in the United States. Nicholas Herbemont would be pleased. Born in France in 1771, as a young man Herbemont immigrated to America from his native land amid the turmoil of the French Revolution. He eventually settled in Columbia, South Carolina. Skilled in viticulture, Herbemont became a very accomplished winemaker, winning praise and awards for his wines and writing extremely influential essays on the proper culture of grapes. He predicted that one day American wines would be the equal of those of France.[18]

But it was a different sort of essay that Herbemont wrote in 1822. Typical of white people who lived in South Carolina at that time, he had been shocked and frightened that spring, due to the plans of another prominent resident of the Palmetto State. Unlike Herbemont, Denmark Vesey lived in

Charleston, not Columbia. More significantly, although both men were free, Herbemont was white. Vesey was black.

Scholars disagree on the scope of Denmark Vesey's planned insurrection. The most dramatic accounts state that Vesey and enslaved followers intended to seize the U.S. arsenal and ships at Charleston Harbor, kill all white inhabitants of the city, and plant explosives to burn Charleston to its foundations. Having destroyed the world of their captors, the rebels would then sail to freedom in Haiti or Africa.[19] Others have suggested that the scope of Vesey's plot was exaggerated by his accusers.[20] In any case, some level of violence was envisioned by Vesey and his followers, but their plans unraveled as African Americans with knowledge of Vesey's scheme informed local whites. Vesey and seventy-seven of his followers were hung or imprisoned.[21]

Particularly discomfiting to the European Americans of South Carolina was their perception that Vesey had no reason for discontent and thus no cause for formulating such a violent plan. Not only had he been free for twenty-two years (having purchased his own freedom thanks to a winning lottery ticket), he was a prosperous carpenter with a comfortable home. He was doing so well that he even employed others to work in his shop. But although Vesey had good relations with his white neighbors and customers, he was deeply disturbed by the status imposed on African Americans of his time. Being literate, he regularly read abolitionist pamphlets. He even perused transcripts of the debates in Congress on whether to admit Missouri to the union as a slave state. It was very clear to Vesey: he could not stand for the status quo.[22]

The U.S. Census figures for South Carolina in 1820, two years before the incident in Charleston, showed that the state had a population of 231,812 white people. These people of European descent were outnumbered by those of African ancestry: South Carolina had 251,778 enslaved people plus 6,714 "free colored persons."[23] Mindful of these numbers, after the planned rebellion was thwarted, Nicholas Herbemont published a pamphlet proposing that South Carolina work hard to attract more people of European descent. To do this, he argued, South Carolina needed to increase its number of sheep, but first it had to reduce its population of enslaved people, perhaps freeing them and then requiring that they leave the state at once.[24]

"Slavery," Herbemont asserted, "is an obstacle to improvements and to the increase of the white population." He added, "[Slavery] is attended with considerable danger." South Carolinians should, he opined, "check the growing evil" or "modify it" to "prevent, if possible, the recurrence of the late disturbances."[25] Herbemont then spent several paragraphs arguing that the enslaved people were well cared for and that their burdens were light.[26] Denmark Vesey's followers would no doubt disagree.

Herbemont next made his recommendations, advocating that South Carolina increase its production of wine, olive oil, silk, and wool to make the state inviting to immigrants to of European descent.[27] Where the wine was concerned, we can grant that Herbemont was selfless in proposing that he should have more competition in his own business. As for olive oil and silk, these had already been tried in the coastal Southeast without much success.[28]

It is the wool that concerns us here. Herbemont devotes only two paragraphs to sheep, and he is very summary in his treatment. Without giving any analysis, he simply insists that sheep would do well in South Carolina and that the demand for wool gave "the greatest encouragement to raise this most invaluable animal." Furthermore, sheep carcasses "would also be a most valuable addition to our meat markets." Making it clear that he wanted South Carolina's shepherds to be white, Herbemont insisted that shepherds "could be easily procured from Europe and kept here at a very small expense." In conclusion, Herbemont admitted that expansion of sheep flocks would not by itself do much to increase the state's human population, "as very few persons are sufficient for a great number of sheep"—again, an emphasis on wool's low labor requirement compared to other agricultural commodities. But even if the employment sheep provided was minimal, he asserted, expanded flocks would still raise white immigration indirectly by increasing the state's wealth.[29]

These prayers went unheeded. The earliest data on the number of sheep in South Carolina is in the 1840 U.S. Census, which reported the state's flock at nearly 233,000 animals.[30] Twenty years later, at the dawn of the war that would end slavery, South Carolina reported 233,500 sheep.[31]

What happened was the opposite of Herbemont's wish. The increase in sheep was negligible, but the expansion of slavery was dramatic. Far from his suggestion that African Americans in the state be emancipated and relocated elsewhere, the enslaved population rose from slightly over 250,000 in 1820 to nearly 600,000 in 1840.[32]

While Herbemont in the 1820s viewed slavery as an evil to be mitigated or even eliminated, it would not be long before the winds would shift.

Assertions that slavery was a noble institution—and outright advocacy for its expansion—would soon predominate in the South. Once again there would be a sheep connection.

⁂

Traveling through Georgia for a week in 1847, Daniel Lee was upset by what he saw. Or, rather, by what he did not see—sheep. He spotted only one. Worse, he saw lots of dogs. Lee was one of a long list of Americans enraged at the high numbers of these animals that they saw as nothing but sheep killers.[33]

Like Herbemont, Lee was a transplant to the South; unlike Herbemont, he was born in the United States. Raised on a farm in central New York, Lee became thoroughly immersed in the science and practice of agriculture, writing and lecturing extensively on how to improve farming practices. He was an advocate of the establishment of agricultural colleges to disseminate knowledge, and accordingly, in 1854, he became the first professor of agriculture at the University of Georgia. By that time Lee had lived in Georgia for several years, having moved to Augusta in 1847 to serve as editor of the *Southern Cultivator*, one of the leading agricultural magazines in the South.[34]

Daniel Lee was also an avid defender of slavery. It had been part of his life even on the New York farm where he grew up, as his father owned several slaves. In 1859 Lee penned several articles for the *Augusta Weekly Constitutionalist* in which he brought together his desire that the South have more sheep and his belief that slavery was beneficial for people, white and black alike. Lee argued that the South could support one hundred million sheep. He postulated that these would need to be looked after by about five hundred thousand shepherds, one person for every two hundred animals in a flock. Additionally, under Lee's plan there would be another one hundred thousand supervisors over the shepherds. Those foremen would be white, but the half-million regular shepherds would be black—and they would have to be brought over from Africa. Just two years prior to the Confederacy firing on Fort Sumter, Lee was proposing that the transatlantic slave trade be revived so that the South would have a sufficient supply of African Americans to take care of sheep.[35]

There were considerable logistical problems with Lee's scheme. The 1860 census, conducted the year after Lee's proposal, showed that the entire United States, including the territories, had nearly twenty-two and a half million sheep.³⁶ Lee was promoting the idea that there could be almost four and a half times more sheep than this in the South alone. E. Merton Coulter's biography of Lee, while sympathetic toward its subject, admits that in this case "Lee was letting his enthusiasm for sheep husbandry run away with his practical sense."³⁷ Coulter offers as a mild excuse that Lee was thinking long-term, perhaps as long as one hundred years. But at no point in the next century would the United States come anywhere close to having one hundred million sheep, let alone having that many in one region.

The proposals to increase sheep promulgated by Nicholas Herbemont and Daniel Lee have similarities. Both men were adamant that the South needed to raise more of these livestock, and both related their plans to the circumstances of enslaved African Americans in the region. But there was also a striking contrast: while Herbemont supported removal of blacks from South Carolina and having an exclusively white workforce of shepherds, Lee desired the opposite—a labor force of shepherds who were primarily African American, coupled with forced removal of Africans from their native lands to be brought into the South to meet the quota he envisioned.

Historians have written of the major change that occurred in the American South regarding the attitude toward slavery. The philosophy of the first several decades after independence was that slavery was wrong—as Nicholas Herbemont put it, "a great evil entailed upon us by our ancestors." But by the 1830s, this thinking became supplanted by an aggressive defense of slavery as a positive good.³⁸

We can see this change clearly by contrasting Herbemont's 1822 proposal with Lee's similar plan thirty-seven years later. When Herbemont mused about the possibility of emancipating the enslaved, this was scarcely an original thought. Thomas Jefferson's *Notes on the State of Virginia*, written during the American Revolution, briefly sketched a plan to gradually emancipate slaves and then recolonize them elsewhere.³⁹ More thoroughly and ambitiously, St. George Tucker, professor of law at William and Mary College, in 1796 published "On the State of Slavery in Virginia," a long essay proposing an intricate system by which all Virginia's enslaved could be freed.⁴⁰

After the 1830s, such thoughts in the South were dismissed as dangerous nonsense. Slavery was endorsed and its expansion desired—as we see in Daniel Lee's idea of enslaving more Africans to guard the sheep. Southern states even criminalized writings or speech advocating abolition or denying a right to property in slaves.⁴¹ From that point until the end of the Civil

War, anyone like Nicholas Herbemont writing that the South needed more sheep would have to do so without concurrently arguing that the region also needed fewer slaves.

Oddly enough, in lamenting that the South lacked sufficient flocks of sheep, white men like Herbemont and Daniel Lee, with their respective excuses and endorsements of slavery, made common cause with Charles Ball, who wrote so extensively of the cruelties of his enslavement. Ball bemoaned: "It is unfortunate for the slaves, that in a tobacco or cotton growing country, no attention whatever is paid to the rearing of sheep—consequently, there is no wool to make winter clothes for the people, and oftentimes they suffer, excessively, from the cold; whereas, if their masters kept a good flock of sheep to supply them with wool, they could easily spin and weave in their cabins, a sufficiency of cloth to clothe them comfortably."[42]

As we have already seen, Ball was correct in his assessment that the South did not have enough sheep to be self-sufficient in its wool needs, although to write that sheep rearing in the South received "no attention whatever" was an overstatement.

John the Shepherd—I will call him that for a bit—was acclaimed in the 1840s as the finest breeder of sheep in Ohio and one of the best in the entire nation. The editor of the *Ohio Cultivator* admitted that when he first read glowing reports in local newspapers praising the wool from John's sheep he was dubious—no wool could be as good as what was gushingly described, he thought. Then he acquired samples from John's flock of Saxon sheep, and he admitted that the quality was indeed superb. Accolades came from out of state as well; a purchaser and manufacturer in Lowell, Massachusetts, said he was "amazed" by John's wool, and he predicted that soon John would be blessed with a "better breed of sheep than are now in existence." When John's wool was displayed at exhibitions in New York City and Boston it was awarded gold medals. It was said that the key to John's success was not only his selective breeding of sheep but also his intensive attention to the health of his flock and to the washing and packing of their wool. There was no question: John could boast of being an expert in sheep husbandry and wool culture.

Books about John—and there are more than several—understandably are less concerned with his sheep-rearing skills than with other aspects of his life, especially his extreme antislavery sentiments and his leadership of the October 1859 raid on the U.S. arsenal at Harper's Ferry, Virginia. John the Shepherd was the abolitionist John Brown.[43]

In David S. Reynolds's extensive biography, the details of John Brown's ovine effort are mostly covered in a chapter titled "The Pauper." This is appropriate, because, while Brown was good with sheep, he was bad at the business of selling their wool. His financial transactions registered few gains and crushing losses. Accordingly, he was constantly broke and trying to stay a step ahead of his creditors. In 1842 Brown's finances had sunk so badly that he declared bankruptcy. Most of his possessions were seized, but he was allowed to keep some farm animals, including seven adult sheep and three lambs.[44]

While John Brown the shepherd was minding his sheep, John Brown the activist was part of the Underground Railroad, regularly hiding escaped slaves as they passed through Akron.[45] It is easy to conceive that some of those he helped escape bondage might themselves have tended sheep in the South. Imagine a fleeing slave, learning that this man supporting his path to freedom, was, like him, a shepherd. What conversations they might have had, swapping husbandry tips or telling stories about their flocks. Perhaps one of those assisted by John Brown told of a dream that one day as a free man he would again take care of sheep—but this time they would be his own livestock, not property of an enslaver.

Reynolds's biography sums up John Brown's entrepreneurial shortcomings: "He failed as a tanner, a shepherd, a cattle trader, a horse breeder, a lumber dealer, a real estate speculator, and a wool distributor."[46] Notice that two of the items in this list of flops were related to sheep—but there is a distinction in how well Brown performed in those capacities. While Brown's efforts as a wool salesman were a debacle, clearly he did not fail as a shepherd, someone in a hands-on role caring for sheep and working to improve their wool. All that praise lavished on his sheep in the 1840s, plus those medals awarded for the quality of their wool, belies any notion that he wasn't highly competent in flock management. He simply was lost once the sheep were sheared and their fleeces went to market.

One might ask whether Brown's life would have turned out differently—or even if the course of American history been altered—had he been as successful in the business of selling wool as in raising the sheep that bore it. Would prosperity as a wool merchant have preoccupied him so much as to lead him to find less extreme means of expressing his antislavery sentiments?

Historian Daegan Miller argues that John Brown was "one of those rare characters, perhaps a world-historical figure in whom is embodied a historical flash point, a point of no return."[47] If Miller is correct, then Harper's Ferry would have happened whether Brown was a superb businessman or, as it turned out, a dreadful one.

One thing is certain. If George Washington, Thomas Jefferson, and the other slaveholding founders of the United States had still been alive in the 1840s, and if they knew nothing of John Brown's politics, nothing of his extreme antislavery actions, nothing of him at all except that he was furiously working to improve America's sheep, those slaveholders would have considered Brown a patriot.

With the defeat of the South in the Civil War and the passage of the Thirteenth Amendment to the Constitution of the United States, the age of enslavement would end. But the ensuing years of Reconstruction and post-Reconstruction would continue to post challenges for the freed African Americans in all aspects of their lives. Their standing within the orbit of America's sheep economy would not be an exception.

John Lord Hayes of Boston was not the type to remain idle in his seventies while suffering from illness that would soon take his life. Rather, he saw his incapacitation as an opportunity to translate Latin hymns of the Middle Ages into English, taking special care to preserve the measure and rhythm of the originals, and then to publish his efforts in a book.[48] Such an effort was true to Hayes's character. Over his lifetime he developed a variety of skills and interests, being an attorney, a successful businessman, and a political organizer. During the Civil War, President Abraham Lincoln appointed Hayes the chief clerk of the U.S. Patent Office.[49]

Also, like so many prominent nineteenth-century Americans, Hayes had a sheep connection, in his case a very strong one. He was the first secretary of the National Association of Wool Manufacturers. This was a duty he took seriously; at the organization's first annual meeting in 1865, he gave a thorough report on the history and current status of the wool business.[50] He published a number of other articles, reports, and books on the topic. If you had a question about wool and sheep in America from the 1860s through the 1880s, Hayes was the expert to seek. In 1877 thirteen governors, U.S. senators, and members of congress from southern states did just that.

The leader of these southern statesmen was Alexander Stephens, the former vice president of the Confederate States of America. John Lord Hayes's brother, General Joseph Hayes, had commanded the Eighteenth Massachusetts Infantry at the Battle of the Wilderness and been severely wounded in the fighting, but it had been over a dozen years since the guns of the Civil War fell silent, and Reconstruction was beginning to fizzle, so it must have been in a spirit of reconciliation that Stephens and the others requested from a Yankee lawyer a report on sheep husbandry and wool production in the South. Perhaps those southern men might have found someone from their own region suited to the task, but as they wrote in their invitation to Hayes, they were impressed with his previous articles. "[In the South, with its special circumstance, these] have not been recently presented by any influential authority, like you represent," they said.[51]

Thus praised, Hayes dutifully got to work on the task, producing a forty-eight-page document, plus a sixteen-page appendix of letters sent to him by southern shepherds. The work also included a section titled "Sheep Husbandry by the Colored Population."[52]

Hayes's treatment of this topic began with a caution that the Department of Agriculture often received complaints of two serious obstacles to keeping sheep in the South: "the destruction of sheep by dogs, and the depredations of lawless negroes." Hayes did not challenge the assertion that freedmen disproportionately pilfered livestock. Instead, he insisted that education would solve the problem. Properly instructed African Americans, he argued, would raise sheep instead of stealing them. Part of that education was economic. "Let the colored people of the South have the means pointed out to them for their *material* improvement," he emphasized. He was certain how this should be done: "What means so simple and ready as the encouragement of sheep-growing among these people, on a moderate scale, in the rural districts?"[53]

Hayes calculated that since there were approximately four million African Americans in the South, this would indicate that there were around four hundred thousand families. If each family had six sheep, that would make 2.4 million wool- and mutton-producing livestock—more sheep than there were in his own New England! This would be a boon to America's wealth, but Hayes declared it would lead to something far more important, asserting "the civilizing and humanizing influence of the pastoral occupation upon the population and the habits of thrift which it would engender." Then, asserting biological support of his argument, Hayes wrote: "The colored race, from their natural gentleness, take most kindly to the care of animals. Negroes, it is well known, make excellent shepherds, as they make capital hostlers."[54]

The stereotypical nature of that comment, along with Hayes's insistence that having sheep would "civilize and humanize" the recently freed blacks, sounds quite racist by the standards of our time. Arguably, though, it was more paternalistic than racist. It was a standard position in eighteenth- and nineteenth-century thought that shepherding civilized *every* society. The thinking was that even the ancestors of a white Boston lawyer who could translate hymns from Latin, like Hayes, had been savages until they adopted a pastoral lifestyle.

Hayes's casual assertion that African Americans were superb shepherds suggests that this represented conventional wisdom among elite whites. Perhaps some of them agreed with Hayes that the new course for blacks in the South was to raise flocks of sheep. African Americans attempting this, however, would encounter some very rough going—assuming in the first place that they could avoid the trap of sharecropping. Most basically, how would the freed people pay for sheep? Hayes optimistically wrote, "There are but few colored families which could not afford to purchase two or three ewes."[55] But, in fact, under the prevailing systems of land tenancy and sharecropping in the South, often African American farmers did not have the capital to purchase livestock.[56]

As noted, Hayes felt sure there were two predators plaguing shepherds in the South: lawbreaking African Americans and dogs. Education, he argued, would turn the freemen into law-abiding citizens. But what was the answer to those marauding dogs? Decades earlier, Nicholas Herbemont had offered a solution. In his 1822 proposal to increase South Carolina's white population by promoting sheep, he had recommended guns: "Shepherds may have loaded fire arms to kill all [wolves and dogs] that approach."[57] Over twenty-five years later South Carolina congressman Richard Franklin Simpson agreed and added more dogs to the solution, writing, "[A] trusty shepherd, with a [trained] dog or two and a rifle, would prevent [canine incursions on a sheep flock]."[58] Thus, if an African American in the post–Civil War South wished to be an effective shepherd, he would be well advised to follow the advice of Congressman Simpson—who in 1860 had voted in favor of South Carolina's secession—and keep a working firearm on hand to protect his charges. Alas, this would prove problematic.[59]

Within several months of Robert E. Lee's surrender at Appomattox ending the Civil War, southern states passed laws designed to keep guns out of the hands of the newly freed African Americans. Mississippi's law declared that no "freedman, free negro, or mulatto" not in military service of the United States was permitted to keep or carry firearms unless he was licensed to do so by his or her county of residence. Florida's law of the same period was more stringent; it required any person of color wishing to keep a firearm to first secure a license from the probate judge of his county, but the judge could only issue the license "upon the recommendation of two respectable citizens of the county, certifying to the peaceful and orderly character of the applicant."[60] Most severe of all was the Alabama statute, which simply declared: "It shall not be lawful for any freedman, mulatto, or free person of color in this State, to own fire-arms, or carry about his person a pistol or other deadly weapon." Here was no military service exception, no way for an African American, whether a shepherd or anything else, to legally own a gun.[61]

Could southern states enact such restrictive and racially based firearms legislation without running afoul of the Constitution? This question came before the U.S. Supreme Court in 1875, in the aftermath of the bloodiest day of Reconstruction. Following a contested and largely fraudulent 1872 gubernatorial election in Louisiana, white supremacist groups seized the Colfax courthouse in Grant Parrish. A militia consisting of freed African Americans evicted them—but the evictees gathered reinforcements for an attempt to retake the building. On April 13, 1873—Easter Sunday—a force of at least 250 men, with armament including a six-pound cannon, attacked the courthouse, defended by about 150 freedmen. Outnumbered and outgunned, the African American militia fell. How many of them died in actual battle and how many were massacred after their surrender is unclear, but the freedmen suffered at least sixty-two fatalities.[62]

Gruesome though their actions were, prosecution of the men who stormed the Colfax courthouse would have been entirely at the discretion of the State of Louisiana had it occurred just a few years earlier. But in 1868, the Fourteenth Amendment to the Constitution of the United States had been ratified. Section 1 of the amendment read, in part, "No State shall . . . deprive any person of life, liberty, or property, without due process of law." Using one of the enforcement acts passed by Congress in the wake of the new amendment's ratification, the federal government initiated legal proceedings against the Colfax rebels, indicting them for conspiracy. The prosecution consisted of thirty-two counts, but only two relate to the story here; these accused the white supremacists of conspiring to prevent the African Americans defending the courthouse from exercising their "right to keep and bear arms for a lawful purpose."[63]

The federal government lost the case in circuit court, and in *United States v. Cruikshank* it lost again on appeal to the Supreme Court. Scholarship on the case typically emphasizes that the courts focused on the words "No State shall" in the Fourteenth Amendment, which the justices took to mean that the Colfax rebels could not be prosecuted under federal law, since they acted as individuals and not as an arm of the state.[64]

What concerns us here, however, is the inclusion in the federal brief of an assertion that the Colfax victims had a right to have firearms "for a lawful purpose." This would include defense against white supremacists intent on doing them harm. But there were other lawful reasons why an African American in the Reconstruction South might want to be armed—such as, for instance, if he simply desired a gun to keep hungry predators away from his sheep. The federal prosecution said nothing in its *Cruikshank* brief about shepherds needing a good rifle; this would not have been relevant to the specifics of the case. But pleading the plight of a freed shepherd would be rendered moot anyway; the Supreme Court's decision validated all state laws restricting African American access to firearms, regardless of the intended use of the weapon. According to the majority opinion, the Second Amendment, although declaring a "right of the people to keep and bear Arms" that "shall not be infringed," applied only to Congress, not to the states. Thus, if African Americans felt that their state was infringing upon their right to a firearm, they would have to appeal to the state for relief—the very state that enacted the law disarming them in the first place.[65]

By focusing on the specifics of the *Cruikshank* decision—and on its origin, the carnage at Colfax—it is natural to frame the racially based firearms legislation in the nineteenth-century South as cruelly leaving freedmen at the mercy of the Ku Klux Klan and other such groups.[66] But the assertions of Nicholas Herbemont, Congressman Simpson, and others that shepherds needed rifles to earn a livelihood shows that this was a far broader issue than suggested by the Colfax violence. Even if a freedman's two-footed neighbors were not aggressive racists, he still would be at an economic disadvantage if he had no gun when confronted with four-footed threats to his sheep.

Another potential setback for the Reconstruction African American shepherd was the so-called closing of the range. Throughout the antebellum period, as a legal matter, owners of cultivated land were required to erect and maintain fences if they wished to keep other farmers' livestock out of their crops. Some of the standards for these fences were quite strict. Georgia, for instance, specified that no farmer could recover damage done to his crops by his neighbor's animals unless he maintained a fence in good shape that was at least six feet high.[67]

This began to change after the Civil War. Gradually over the next eighty-five years the South moved from "open range," where the crop owner had to maintain a fence to keep livestock out, to "closed range," where the process was reversed—the livestock owner had to maintain a fence to keep his animals in. Some of the reasons for this are beyond the scope of this study, but one significant catalyst for the change was the presence of formerly enslaved African Americans, now free. Where livestock ran at large, they were vulnerable to being caught and slaughtered for food by someone other than the owner. Whether the pilferer was white or black, the blame often fell on the former slaves. "We cannot raise a turkey, chicken or hog in consequence of the stealing by negroes," attorney and former Confederate general James Holt Clanton complained, speaking specifically of the situation around Montgomery, Alabama. Laws to confine livestock would thus not only protect crops but also restrict the places where these animals could wander, perhaps reducing the chance they could meet their end at the hands of a hungry neighbor or vagabond.[68]

But, more significantly, such closed-range laws would also be unfavorable to a freed African American planning to heed John Lord Hayes's advice to mind one's own "material improvement" by becoming a shepherd. Rather than being allowed to let his sheep to range as they willed, as had been the custom in the antebellum South, he would now have to build and maintain fences to enclose his flock—a problem not simply of obtaining capital to do so but sometimes also one of scarcity of supply, given timber shortages in some areas.[69]

Moreover, in examining the parts of the Deep South where fence laws were first implemented, historian J. Crawford King determined that demographics played a major role. Instead of statewide legislation, Alabama and Mississippi delegated to the counties the authority to determine whether all or part of their land should be open or closed range—and, as King discovered, "the counties that first abolished free range were those with the overwhelming majority of blacks." A former slave in these places envisioning a jump into sheep raising would first have to engage in some fence building, a formidable burden to his start-up plans.[70]

Even legislation that on its face seemed perhaps designed to protect the freed African Americans might potentially inhibit any plans for extensive sheep culture. For example, in 1866 North Carolina passed a law declaring invalid any contract for the sale of livestock—including sheep—in which one or more of the parties was a "person of color" and the purchase price was ten dollars or more, unless the contract was in writing and witnessed by a literate white person. This could plausibly be defended as good consumer protection—but it is quite easy to see how a freed African American's suspicious or unsympathetic neighbors could curtail any plans he had to start or increase a sheep flock by not giving their signed assent.[71]

African American shepherds of the post–Civil War years faced still other challenges. Nineteenth-century Americans regularly touted the value of sheepdogs to the shepherd, both to herd the flocks and to keep at bay wolves or feral dogs that might otherwise kill sheep. During and after Reconstruction, the taxes southern states imposed on keeping dogs probably affected African Americans more than whites.[72] This is discussed in the next chapter.

How much did these potential roadblocks actually prevent postbellum African Americans in the South from becoming shepherds? In one sense, it is difficult to determine. If a freedman contemplated acquiring sheep, but his inability to also purchase a firearm, pay a dog tax, or build fences prevented him from succeeding in the endeavor, his unfortunate experience would likely go unrecorded.

There is, however, some concrete data, thanks to the work of the noted African American writer and civil rights activist W. E. B. Du Bois. A 1900 census bulletin on the status of African Americans contained a chapter written by Du Bois titled "The Negro Farmer." He used the phrase "farms operated by Negroes," since he included both farms owned by African Americans as well as those operated on behalf of others. Thus his statistics do not necessarily connote ownership.[73]

Du Bois counted only 5,672 sheep-holding farms operated by African Americans. Collectively these farms held 97,550 sheep, making an average flock of slightly more than seventeen animals per property.[74] Thus there

Shepherds Enslaved and Free 73

were far fewer shepherds of color than the four hundred thousand African American families each with six sheep that John Lord Hayes had envisioned two decades earlier. One aspect of Hayes's proposal did take root, however—its regional basis. Du Bois reported that over 80 percent of U.S. sheep managed by African Americans were on farms in the South, showing that, in spite of the obstacles, some freedmen did become shepherds.[75]

One was Scott Bond, born into slavery in Mississippi. He was described in a glowing 1917 biography as "one of the largest land owners, merchants and stock-raisers in Arkansas." Bond did not complain of any racially based impediments to his establishment of a sheep flock. He remarked that he paid fifty dollars for a fine ram at a state fair and that he thought the price quite high, but he wound up happy to have made the purchase. The ram serviced his ewes, who proceeded to give birth to offspring that produced three and a half pounds of wool, a big improvement over the two pounds had he harvested from their mothers. Armed with this personal experience and with an admiration for ovine adaptability, Bond insisted that sheep culture was good economics. "Besides," he playfully added, "it is a real pleasure to look at a flock of sheep and to watch the lambs as they gambol in the springtime."[76] Like the Wayman family we met at the start of this chapter, Bond was proud to have his freedom and his sheep.

CHAPTER 5

Fido and the Fleece

Sheep, Dogs, and the Law

∽∂∂∽

> Never before did we ever hear of Reverend D. W. Coldwell, of this county, getting mad until last week. Dogs killing four of his sheep was the cause.
> —*Dahlonega (Ga.) Nugget,* June 25, 1909

Even a man of God has limits to his patience. Sheep appear frequently in the Bible preached by Reverend D. W. Coldwell.[1] But those who owned flocks, even if they were Christian clergy like Coldwell, regularly cursed dogs as though they deserved more revile than the serpent in Genesis. For much of American history, people raising sheep complained bitterly of dogs massacring their flocks.

Not far from where Coldwell lost livestock to dogs in 1909, the Millers of Floyd County, Georgia, in 1908 had similar problems. A dog belonging to the Millers' neighbors regularly wandered onto their property. On these forays, the trespassing dog had killed at least eight of their sheep and injured several more. Mr. Miller approached Mr. Stanton, the patriarch of the neighboring family, and made the not unreasonable request that he keep his destructive animal off Miller's land. Stanton replied that he had no intention of confining the animal. Accordingly, Miller decided that the next time the dog attacked one of his sheep the canine would be on the receiving end of a bullet. He shared that decision with his young son, giving the boy permission to take down the four-legged marauder if circumstances warranted.

Obviously, the Stanton's dog was not going to stop being destructive just because words were exchanged among humans, and, true to form one night, the canine killed more of the Miller's sheep. The next day the Miller boy took his weapon, walked over to the Stanton's land, and—in what was later called "a serious breach of propriety and lack of neighborly consideration"—shot the dog to death right in front of the Stanton family.

Young Mr. Miller was subsequently tried in Floyd County court for a misdemeanor—cruelty to animals. In the charge given to the jury, the judge told them that Miller had "no right to kill the dog . . . even on his own premises, unless it was caught in the act of killing or interfering with the defendant's sheep." Given that the defendant admitted that he hadn't killed the dog while it was attacking a sheep, but later after the damage had been done, the jury convicted him. Miller appealed, and in the 1909 case of *Miller v. State*, the Georgia Court of Appeals reversed the judgment.[2]

Judge Arthur Gray Powell showed his sympathy for dog owners, but as a matter of law he ruled that affinity for canines was secondary to the property rights of sheep owners. The judge noted, "[In some states] the sheep-killing dog is made an outlaw by statute, [but in Georgia] his status is a part of the higher or unwritten law." Powell acknowledged that if a dog was guilty of "the worrying of sheep" just once, it might not be justifiable to kill the canine, "but when a dog acquires the . . . sheep-killing habit, he becomes a nuisance and may be destroyed as such."[3]

Getting to the specifics of the case, the judge countered the notion that the Stanton dog could only be shot when caught in the act. Sheep-killing dogs were "so sly and wary when engaged in their nefarious practices," Powell reasoned, that they "elude every approach of the owner of the sheep." If the sheep owner could only shoot dogs caught in the act of killing his livestock, this would provide the shepherd "a very inadequate protection."[4]

Judge Powell did not absolve Miller of the animal cruelty charges; he simply declared that whether the Stanton dog was killed in a spirit of cruelty or justifiably shot as a "menace to property more valuable than itself" were questions for a jury, one charged under new instructions that it wasn't necessary to observe a dog in the act of killing sheep to be within one's legal right to kill it.[5]

Perhaps Powell feared that his decision would appear callous. This would explain why he concluded his opinion by noting the fondness people—including himself—have for their dogs:

> [The Stantons] doubtless loved the little fice; these little animals, however worthless they may be, have a way of endearing themselves, especially to the women and children of the family. I well remember how in the days gone by my childish tears flowed as, in poignant grief, I stood broken-hearted and

viewed the cold remains of my fice dog, Buster, who had met an untimely death. But under the record we are inclined to think that the defendant's cruelty was operative against Mr. Stanton's family, rather than against the dog, which seems to have been worthless and of a vicious temperament. He did wrong to shoot when and where he did, but he is entitled to a new trial as to the penal offense with which he stands charged.[6]

Dogs are so beloved and ubiquitous in the United States today—and sheep are comparatively so scarce—that it may be jarring to hear that in the past many strongly asserted that ovines took precedence over canines. But this they did, and conflicts between shepherds and dog owners were once a regular feature of American life. This chapter examines that frequent clash. *Miller v. State* is a good starting point for several reasons. First, note how strongly Judge Powell viewed the legal standing of a sheep-killing dog as a "nuisance" animal that could be destroyed with impunity—he asserted that this was a matter of "higher law" even in the absence of a relevant statute. Second, in spite of his pro-sheep ruling, Powell acknowledged that there are strong human affections toward dogs, no matter, as he put it, "how worthless they may be." This friction between the economic value of sheep and the affective value of dogs would undergo modification as America became more urbanized, leading to the increase of canine-loving city dwellers separated from the rural source of their livestock-based goods. That the Miller case was decided in 1909 shows how late dog-versus-sheep squabbles persisted, and, indeed, we will see that sheep still enjoy some legal preference over dogs today.

At Oxford University in 1753, Sir William Blackstone delivered a series of lectures on English law, later published in four volumes between 1756 and 1769. Titled *Commentaries on the Laws of England*, the series sold well in Britain and nearly as well in the American colonies soon to break away. The books were widely read by the revolutionary generation and remained instrumental to American legal thought well into the nineteenth century. No doubt Judge Powell was familiar with the books.[7]

In a chapter on property published in the second volume, Blackstone noted that animals, like any chattels, can be stolen by unscrupulous individuals.[8] But under the common law the severity of the theft depended on the type of animal pilfered. If the stolen animal was a horse, cow, sheep, or chicken, the human transgressor had committed a felony because "these are things of intrinsic value, serving for the food of man, or else for the uses of husbandry." Not so if a dog was taken, in which case, stated Blackstone, that was merely "a civil injury . . . redressed by a civil action." This, he wrote, was because dogs (and cats) have no intrinsic value: "They are only kept for pleasure, curiosity, or whim," their worth "depending only on the caprice of the owner."[9] Dog and cat lovers of today might squirm at the notion of their beloved companions legally occupying a rung below that of sheep and chickens. As seen in *Miller v. State*, however, Judge Powell had similarly argued that even though people may love their dogs, these animals are basically worthless—a less tactful way of putting what Blackstone had written a century and a half earlier.

But there were powerful currents—or perhaps we should say powerful bonds of affection—serving to undermine the rigidity of the common law view. Indeed, Blackstone may have been describing a situation that existed *legally* in the England of his time, but it scarcely applied socially. By the late eighteenth century, there were thought to be almost a million dogs in England, and most of them were kept for pleasure rather than for need.[10]

Use of dogs as working animals persisted, however, and the most revered of these—in England, the United States, or anywhere else shepherds kept their flocks—were breeds that assisted in herding sheep.[11] Thus, to the sheep owner, a dog could be one of two things. It could be an ally that made sheep maintenance simpler and more efficient. Or, as Reverend Coldwell and Mr. Miller could attest, it could be a nightmare, killing valuable livestock. America's sheep farmers would come to be quite vocal about the damage they suffered from dogs.

"For some time hence this will not be a great sheep country," Judge Richard Peters predicted of his home state of Pennsylvania when he wrote George Washington in 1792. "We keep too many dogs who destroy them."[12] Peters was too pessimistic; just eighteen years later Pennsylvania had over 618,000 sheep, far more than any other state that reported its ovine population.[13]

But even with the increase of Pennsylvania's flocks, Judge Peters continued to urge vigilance toward dogs. In 1810, as president of the Philadelphia Society for Promoting Agriculture, he detailed the problem in an address to the organization. Peters lamented that recently his shepherd had carelessly left his prized flock of Tunis sheep out in the field instead of penning them in the fold; as a result, they had been attacked by dogs. "The flagitious sagacity of dogs is almost incredible, when they are addicted to sheep killing," Peters complained. "Of this vice, when it is once fixed, they are never cured while living: death is the only effectual remedy." He shared a story reported to him by a Maryland acquaintance who said that dogs had killed 130 sheep in his neighborhood in ten days. An elaborate trap was set, basically a pen designed so dogs could enter but not escape. It was baited, and in one night seven dogs were caught, a point Peters emphasized to assure his audience that dogs were especially destructive since they prowled in packs and killed more sheep than were needed for their nourishment.[14]

Peters's address acknowledged that the blame lay not with the dogs themselves but rather with their human masters, and he showed some concern for canine welfare. Many dogs were useful as protectors of property—including sheep, Peters admitted. The problem arose because dogs "are kept in too great numbers, and of breeds, in many instances, worthless." When dogs were not properly cared for, they were "compelled to prowl for sustenance." Peters then unleashed his full wrath: "It should be made disgraceful and *uncivic* in those who keep supernumerary, worthless, or starved dogs. They injure society, by exposing the persons of their fellow-citizens to disease and death, and their property to plunder and destruction, when such dogs become mad, or ravenous *beasts of prey* (emphasis in original)."[15]

Fears that dogs would interfere with America's plans to become a sheep mecca were not confined to Pennsylvania. Indeed, when George Washington received Peters's 1792 correspondence, he no doubt thought how much the situation Peters described mirrored what he experienced in Virginia. Kept away from his farms by his duties as president of the United States, Washington wrote Anthony Whitting, his farm manager, lamenting the regular reports of sheep lost to marauding dogs. To protect his flocks, Washington encouraged his manager to diligently kill any stray canines he could.[16]

Thomas Jefferson voiced similar complaints. Jefferson's passion for sheep improvement in Virginia has been noted in chapter 2, but in 1792 he lamented the "great obstacles to their multiplication"—disease, wolves, and dogs. He observed that wolves were a threat only in those wilder western areas of the Virginia, but domestic dogs preyed on sheep "in all parts of it."[17] Twenty years later, after he had retired from politics, Jefferson elucidated the nature of the love-hate relationship a farmer of his time might have with dogs. They could be "the most watchful & faithful of servants," he acknowledged. But if neglected and not reasonably fed, they would become "the most destructive marauders imaginable." "You will see your flock of sheep & of hogs disappearing from day to day."[18] Two years later Jefferson reported to his brother Randolph that he had caught one of his own dogs in the act of eating a sheep she had killed. "She was immediately hung," Jefferson tersely and unsentimentally wrote.[19]

Given the apprehension Washington and Jefferson had over free people owning dogs, it isn't surprising that they did not want their enslaved people keeping these animals. Both men were determined to keep the slave quarters on their plantations canine-free. In the same letter in which Washington instructed his farm manager to kill stray dogs, he sternly ordered, "If any negro presumes under any pretense whatsoever, to preserve, or bring [a dog] into the family, . . . he shall be severely punished, and the dog hanged." To Washington it wasn't just that an enslaved man's dog might kill sheep, he was convinced any African Americans who kept dogs must be up to no good. "It is not for any good purpose Negros raise, or keep dogs; but to aid them in their night robberies," he told Whitting.[20]

Jefferson's opinion was similar, and he emphasized that his objection to the enslaved having dogs was firmly tied to his concerns about sheep. Writing to his overseer, Edmund Bacon, shortly before ending his second term as president, Jefferson sternly commanded, "To secure wool enough, the negroes dogs must all be killed. Do not spare a single one."[21]

As European Americans from Virginia, the Carolinas, and Georgia moved into the new states of the Old Southwest, they took their enslaved people with them, as well as their sheep. Disapproval of dog ownership by African Americans also moved west, and again the stated reason centered on sheep. Prior to his career as a landscape architect, Frederick Law Olmsted worked as a newspaper correspondent. In the 1850s he traveled through the South, interviewing the region's residents. Noting in one of his reports that an Alabama statute denied dogs to the enslaved, Olmsted wrote that his understanding was that the law "was intended to abate the great destruction of sheep by negroes' dogs." This, he added, was "an evil which is everywhere complained of at the South, and which operates to prevent more extensive wool-growing there."[22] Obviously dogs will kill sheep regardless of their owner's color, so the Alabama act seems more like racism than a serious effort to alleviate the problem.

It was one thing to control canine ownership among an enslaved population through rigid legislation or severe mandates by plantation owners. But no one doubted that many free white people also had sheep-slaying dogs. How could this problem be addressed? A tax on dogs, some thought, was the answer.

Early in 1809, Thomas Jefferson sent wool samples from his sheep—which he identified as Merinos—to one of his regular correspondents, James Ronaldson of Philadelphia. The two men had exchanged letters for several years, sheep and wool being a favorite topic. Ronaldson and his fellow Scottish immigrant, Archibald Binny, had formed the printing partnership of Binny & Ronaldson.[23]

In the letter accompanying his samples, Jefferson asked Ronaldson's help determining the value of the wool enclosed. Jefferson assumed the hatters of Philadelphia would be the best judges of the wool's quality. Could Ronaldson please show the samples to a few of those hatters and report back on their assessment? Showing his competitive nature, Jefferson admitted that he especially wished to know how his wool compared with "that of Colo. Humphreys, Chancelr. Livingston & Mr. Dupont," the only people he know of who had Merinos. Restating his belief that sheep improvement should be on the mind of all good American farmers, Jefferson concluded the letter: "I must pray you to pardon me the trouble I give you, the object of it being to know how far it is worth my while to persevere in the same breed *& to encourage my neighbors in it* (my emphasis)."[24]

Ronaldson graciously replied, first gently correcting Jefferson on his belief that hatters used Merino wool. The long-stapled wool from that breed, Ronaldson assured him, rendered it unfit for that business. Helpfully, Ronaldson then consulted a stocking maker who assessed Jefferson's wool at a value of about forty-five cents a pound.[25]

Ronaldson then shared with the former president news about the sheep-raising efforts he and Binny had undertaken. The two men had long thought of getting sheep as a side business, but they had "been deterred from going into it by the mischief sustained from Dogs." But recently there had been good news on that front, thanks to the Pennsylvania legislature. They had, Ronaldson reported, passed a law taxing dogs: "the proceeds of which tax is to form a fund to indemnify those who suffer from

Fido and the Fleece

the depredations of these animals." Armed with this government assistance, the printer partners figured sheep were a sound investment, one that Ronaldson calculated "must be more profitable than grazing large cattle or cultivating grain."[26]

Two weeks later Ronaldson sent a follow-up letter providing a draft of the dog tax bill he had mentioned in the previous note.[27] That legislation, applicable to Philadelphia and the surrounding counties, commanded the commissioners of those districts to annually census the dogs in their jurisdictions. Based on this data, the commissioners were to assess a tax on the dog owners: twenty-five cents if they had one dog, another dollar if they had two, and two dollars for each additional canine held. The law further provided that most of the money collected from this tax was to be used "as a fund for remunerating the inhabitants of the said counties, respectively, for any loss they [should] sustain after the passage of this act, by sheep being destroyed by a dog or dogs." In other words, it was made abundantly clear that the reason for the new tax was exactly what Ronaldson reported to Jefferson—as economic assistance from government to those sheep owners suffering unsustainable losses from dog predation.[28] Furthermore, if the amount collected by the authorities exceeded the loss claims, the law commanded the excess to be used to purchase full-blooded Merinos. "[In this case], every farmer shall have liberty to send three ewes to [a Merino] ram in said county to continue with him for one week free of expense, except a reasonable compensation for pasturage."[29] One way or another, the drafters of the tax law intended that dog owners were going to help pay for local sheep improvement. The legislation also authorized any person witnessing a dog chasing or attacking a sheep to kill the offending animal—precisely the situation adjudicated in *Miller v. State* a century later.

The movement to tax dogs soon came to Virginia. In 1811 Jefferson received a correspondence from Peter Minor, a fellow Albemarle County farmer. The letter had an enclosure detailing what it called "a law to encourage the raising of sheep." The enclosure has not been found; fortunately in the text of the letter Minor gave a summary: he was passing along a petition to the Virginia legislature for a dog tax. Minor seems to have been the primary author of the petition, but he graciously credited the source of its inspiration. "The principal features," he wrote, "I have taken from the Pennsylvania Dog Law." Minor encouraged Jefferson to "assist . . . in trying to obtain its passage by the next legislature."[30]

Jefferson needed little convincing that Minor's plan was a good one. Within a few days he penned a reply with a rather startling opening sentence. "I participate in all your hostility to dogs, and would readily join in any plan for exterminating the whole race," he declared. "But as total extirpation cannot be hoped for," Jefferson lamented, "let it be partial." The remainder of the letter suggested slight changes to the draft of the petition and concluded with a modest statement: "I know of no service I can render in this business, unless perhaps to write to some friends in the legislature to interest themselves in promoting it."[31] A few months later the petition was presented to the fall 1811 session of the Virginia General Assembly. It bore Jefferson's signature as well as over eighty others. The petition stressed the importance of sheep improvement to the commonwealth, praising those individuals who "at great expense & trouble" had "improved the quality of their wool by the introduction of the best breeds from Spain & other foreign countries." But it decried the dark cloud looming over all those raising sheep: "the ravages they are subject to from Dogs, who in one night may not only devour their property . . . but may damp & perhaps forever destroy all future enterprise & energy in the same cause." The appeal proposed a tax on dog ownership, with the proceeds used primarily to compensate those who lost sheep due to those dreaded canine "ravages." The proposal was thus, as Peter Minor earlier acknowledged, nearly identical to the Pennsylvania act.[32]

After receiving the petition, the Virginia House of Delegates referred it to a committee, which gave a favorable report. A suitable bill was drafted, but it failed to pass. Just two years later, however, in 1814, Virginia enacted a similar law titled "An Act to Prevent the Destruction of Sheep in This Commonwealth." As if the title of the legislation wasn't clear enough to loudly proclaim its purpose, the preamble got right to the point. It began, "Whereas the improvement of the breed of sheep and their numbers in this commonwealth is at all times important but more especially at this time," and then it promulgated the details.[33] Under the law, a Virginian would pay a tax of one dollar on every dog over six months old. If he owned a plantation, he could have two dogs tax-free. For those short on cash, the legislature provided that each wolf or fox scalp submitted amounted to a tax credit good for four dogs. A magistrate finding sufficient evidence that a particular dog had killed sheep could order that dog killed. But unlike the earlier proposal or its Pennsylvania inspiration, the Virginia statute did not mandate that the revenue collected go to sheep owners after a canine massacre of their flocks.

Instead, it directed that funds be used to support poor people. This tax only applied to specific portions of the state; oddly enough, Jefferson and Minor's Albemarle was not one of the counties covered.[34]

※

Regulation of dogs would remain an ongoing process throughout the nineteenth century. The propensity of dogs to kill sheep seems to have always been the catalyst spurring these legal efforts. The dog question was addressed with particular force in *Origin and Growth of Sheep Husbandry in the United States*, an 1879 report by the U.S. Department of Agriculture focusing on the challenges facing shepherds in the South. This document quoted letters from thirty-nine sheep owners, all plaintively lamenting that dogs impaired their best efforts. "Sheep raising has but little attention from our farmers," a Mr. Hammett of southern Georgia declared, "from the fact that if a man gets a good flock started, he does not know what night some cur or hound will kill them all. Sheep-raising can never be successful until we get rid of the dogs." A Mr. Harrill of North Carolina concurred, adding: "Sheep are an unprofitable investment. The annual increase is canceled by the loss from dogs." Writing from northern Mississippi, a Mr. Kimbrough complained, "The advantages of a most bountiful pasturage and good climate are more than canceled by the ravages of the dog."[35] The report estimated that of the nearly ten million sheep in the southern and border states, over five hundred thousand were killed each year by dogs, an annual flock loss of about 5 percent and an annual monetary loss of $1 million. Leaving no doubt where the Department of Agriculture stood, the document melodramatically described the damage done by dogs as "a tax paid by a few promoters of a useful infant industry to encourage the extension of a race of mangy curs too worthless for valuation, yet costing other millions to feed, taking bread from the mouths of the half-fed children of the poor."[36]

Origin and Growth of Sheep Husbandry also noted that sheep were more vulnerable to dogs than other livestock. An excerpted letter from a Mr. Early in Virginia observed: "As the dogs have killed so many sheep, the attention is turned to cattle."[37] Frequently in America there was animosity between sheepmen and cattlemen. If dogs were an equal scourge to cattle and sheep, we might expect that in this area, at least, there would be common cause between the bovine and ovine advocates. But dogs do not pose the same danger to cattle as to sheep, due to the difference in the size of these livestock.

A nineteenth-century adult Hereford might weigh over three thousand pounds, too big for most canines to kill. In contrast, the largest sheep I have taken care of was a male Southdown that tipped the scale at around two hundred pounds, and he was unusually large.[38] An advantage cattle culture had in colonial America was that sheep were more likely to succumb to wolf predation. Dogs too could readily prey on sheep, so this argument in favor of cattle persisted even after wolves were largely extirpated.

The 1879 report concludes its section on dogs with a furious letter from a Mr. A. K. Denny. He, like many other sheep boosters, was convinced that only the enactment of a tax on dog owners would solve the problem, and he offered his opinion on why such laws were difficult to get passed:

> As a State we have not reached that advanced civilization which will promptly give to the sheep, that innocent, defenseless, most useful animal, that protection which its position as a food and clothes producing animal demands. Our legislature very readily passes most stringent laws for the protection of the fish in our waters, the squirrels of our forests, the rabbits in our fields, and all kinds of birds of the forest and field and those that soar above the earth, but when it comes to their best friend, one they cannot possibly do without, they become paralyzed with fear, and proceed with a politician's caution and are careful to pass no law that would wound a voter's feelings or hurt a high-born dog.[39]

A reader from our time encountering these words, in a paragraph referencing both sheep and dogs, could easily be confused, and need to reread it carefully. Wait a minute—Denny is calling the sheep the "best friend" of humans! But isn't the dog that he so loathes actually "man's best friend"? Coincidentally, less than a decade before Denny wrote the letter, a classic expression of a dog's faithfulness toward its master was made by a Missouri lawyer. It happened because of a court case arising from the actions of a man protecting his sheep by shooting dogs.

It was the autumn of 1869, and Leonidas Hornsby of Johnson County, Missouri, was fed up. Dogs had killed many of his sheep—he claimed he lost one hundred of his flock this way. Accordingly, he vowed that he would shoot any dog that set foot on his property. The result of his vow would be a lawsuit.

To this point, Hornsby's story basically matches the experience described earlier of Mr. Miller of Georgia. But while the Georgia case of *Miller v. State* is not well known, the Missouri case of *Burden v. Hornsby* is famous, its renown due entirely to one of Burden's attorneys—a future United States senator—making one of the most memorable closing arguments in the history of American trials.[40]

The trouble began one night in late October when Charles Burden, a neighbor of Hornsby's, heard a shot coming from the direction of the Hornsby homestead. Burden later testified that he immediately was fearful that one of his dogs had been shot, so he blew his hunting horn to summon them to ascertain that they were all right. All his dogs came running in response to the horn, except his prized hound Old Drum. Concerned, the next morning Burden went looking for his special dog.

Arriving at the Hornsby place, Burden asked his neighbor if he had seen Old Drum. Hornsby replied that he had not. Burden then remarked that he had heard the shots the previous night and assumed Hornsby was shooting at a dog. Hornsby admitted that he instructed Dick Ferguson, a young man living with the Hornsbys, to shoot at a trespassing canine, but Hornsby knew what Old Drum looked like, and he insisted that was not the dog fired on. Furthermore, Hornsby maintained, Ferguson's gun had been loaded with dried corn, not bullets, the intent being to scare off an animal rather than to kill it.[41] Burden resumed the search the next morning and found Old Drum's body adjacent to a nearby creek, the dog's head lying in the water. The deceased dog appeared to have been dragged there. Burden was devastated; he filed a lawsuit against Hornsby for the loss of his prized hound.

The legal proceedings were quite protracted, but in the end, Burden won and Hornsby had to compensate him. It seems remarkable, given the evidence, that Hornsby could have been found guilty. As noted, he admitted to instructing Dick Ferguson to shoot at a dog but denied that it was Burden's animal or that the corn shot was fatal. When Burden visited the Hornsby homestead, Ferguson even took him to where he said he shot at the dog, and Burden later admitted there was no blood at the site. When Burden found Old Drum's body he noted that there were sorrel colored hairs on it, apparently from a mule. Burden suspected that these were from the mule used to drag the dog's body. Hornsby owned a sorrel mule—but so what? According to the 1870 census there were more than twenty-one hundred mules and asses in Johnson County.[42] Just in Kingsville Township, where Burden and Hornsby lived, there were fifty landowners who had mules or asses, and there was no telling how many of these were of a brownish color.[43] In the absence of modern forensics, what reason would there be to conclude that the beast of burden involved was Hornsby's mule? In any case, it would be awfully careless for Hornsby to go to the trouble of moving a dead dog to a place where its determined owner could so easily find it. There simply does not seem to be a preponderance of evidence that would lead an impartial jury to find Hornsby at fault.[44]

Lore holds that part of the reason Burden prevailed was the brilliance of one of his lawyers. As the suit passed through the stages of its adjudication, this simple civil proceeding between two neighbors ultimately involved the most distinguished attorneys in Missouri. None were more distinguished than George Graham Vest, a member of Charles Burden's team.

At the time of the Old Drum case, Vest was not quite forty years old. After the Old Drum case he would go on to have a notable career in politics, in 1879 becoming a U.S. senator who would go on to represent Missouri in the upper house of Congress for twenty-four years.[45] In the Senate, Vest was a tireless advocate for preservation of America's unspoiled western lands. In this bipartisan effort, Republican president Theodore Roosevelt came to see the Missouri Democrat as an important ally in his conservation goals.[46] Far less charming than his environmental politics was Vest's vocal opposition to women voting.[47]

In spite of his years and his work in the Senate, however, Vest himself seemed to understand that he would always be better known for what he had done almost a decade before he went to Washington, when on a September day in 1870 he argued on behalf of Charles Burden. Vest arose in the Warrensburg, Missouri, courthouse, faced the six men who would render the verdict, and began to speak:

> Gentlemen of the jury. The best friend a man has in the world may turn against him and become his enemy. His son or daughter whom he has reared with loving care may prove ungrateful. Those who are nearest and dearest to us, those whom we trust with our happiness and our good name, may become traitors to their faith. . . . The one absolutely unselfish friend that a man can have in this selfish world, the one that never deserts him, the one that never proves ungrateful or treacherous, is the dog . . . a man's dog stands by him in prosperity and in poverty, in health and in sickness. He will sleep on the cold ground when the wintry winds blow and the snow drives fiercely, if only he can be near his master's side. He will kiss the hand that has no food to offer, he will lick the wounds and sores that come in encounter with the roughness of the world. He guards the sleep of his pauper master as if he were a prince.[48]

It is uncertain how much of this account is accurate. A brochure published in 1957 by the Warrensburg Chamber of Commerce—people who obviously wished the story presented in as glowing a light as possible—called Vest's plea "The Lost Speech" (it is often today referred to as "Eulogy of the Dog"), admitting that since there was no stenographer present to take down the words, no published version existed until at least 1885, or perhaps not until as late as 1901, there being conflicting stories on when and where someone finally put this tribute to dogs on paper.[49] As such, we really do not know if the words we read today are exactly as Vest expressed to the jury.

Furthermore, the story Edwin French, author of a short biography of Vest, tells of the circumstances that led Vest to make this closing argument is astonishing. French admits that the tale had been related to him by "a member of the Senator's family," but he seems to have accepted it without question. According to the unspecified family member, Vest's law partner, John F. Philips, was the lead attorney in the case. Vest, in contrast, "kept absolutely quiet and took no part in the proceeding." Nor had he planned to address the court, but Philips cautioned him that if he did not do so, their client would object to paying a silent lawyer. Thus it was quickly decided that Vest should make the closing address to the jury. "Young Vest then arose," the account declared, "and without notes or any preparation at all, spoke the few words which have become so famous."[50] It is hard to believe that such a carefully crafted speech as it was later written could be delivered without extensive effort put in beforehand.

In any case, accounts of the trial insist that Vest's argument was delivered with great passion and had the desired effect. One source reports, "When Vest closed, there were few present, the jury included, whose eyes were dry."[51] This cannot be confirmed by court documents. The terse, official record tells us simply: "[The] jury ... after hearing the evidence—handed in the following verdict ... We the jury find for the plaintiff the sum of twenty-five dollars & costs." There is no mention of tears.[52]

But the question inevitably arises: Was George Vest's argument a wise one? Vest's "Eulogy of the Dog" seems like a superb argument for a lawyer in twenty-first-century New York City trying to convince a jury to side with his client whose dog was hit by a delivery van owned by a wealthy corporation. But was it a good close for the prosecution in rural Missouri in 1870, where unlike modern Manhattan there was a strong chance a jury would include jurors who had lost sheep to dogs and thus might not be sympathetic toward Old Drum and his owner?

Thanks to the U.S. Census of 1870, the same year *Burden v. Hornsby* was tried, there is data to better understand potential jury bias. The census tallies include 176 agricultural returns from Kingsville Township of Johnson County, where Charles Burden and Leonidas Hornsby lived. Forty of those 176 respondents reported that they had sheep. Mostly these were small flocks of from six to thirty animals; only three people in the township had more than forty sheep. Two of those three were members of the Hornsby clan—Leonidas reported one hundred sheep, while his father had a flock of two hundred. So we can see that while sheep ownership was common in the township, almost no one else there had as vested an interest in their flocks as the Hornsby men.[53]

While the official record of *Burden v. Hornsby* is sparse, the names of the six jurors trying the case are included. Five of those men also are among the 176 names on those 1870 agricultural returns.[54] Of those five, two of them reported having no sheep, the third man had one sheep, the fourth had ten sheep, and the fifth had a flock of twelve. Thus, where sheep were concerned, the jurors truly represented Kingsville Township: all of them possessed either no sheep or just a small number. No doubt they had some sympathy for Leonidas Hornsby's plight, but certainly they weren't men like him, owners of large flocks regularly ravaged by dogs. In that sense it was not, strictly speaking, a jury of Hornsby's peers.

If George Vest knew any of this, there is no record of it. But it's easy to see circumstances where Vest's strategy could have backfired. Given the reality of 1870s rural America, where dogs were despised in many quarters, it is evident how a jury in a case like *Burden v. Hornsby* could have been comprised largely or totally of sheepmen who would have been unmoved, even irritated, hearing Vest declare that a dog guards the sleep of his pauper master as if he were a prince. Had the panel consisted of men with experiences similar to those of Reverend Coldwell, Richard Peters, or Thomas Jefferson, it seems likely they would have empathized with Hornsby, not Burden, and thus smirked as Vest orated. They might even have scornfully thought, "No, while his pauper master is sleeping the dog won't guard him—it will romp off into the night to kill my sheep!" In this light, Vest was taking a risk.

In any case, when the trial was over and Charles Burden won his twenty-five-dollar settlement, one imagines it might have occurred to him that while a man's best friend might be his dog, when that man goes to court his best friend is a competent attorney.

In 1971, just over a century after *Burden v. Hornsby*, the American novelist John Gardner wrote *Grendel*, a retelling of the *Beowulf* fable from the perspective of the monster.[55] Gardner shifts the point of view of the legend so that, instead of Grendel being the fearsome, bloodthirsty antagonist he is regarded by Hrothgar and his men, he is presented as intelligent and thoughtful and expresses his reasons for attacking the mead hall, an act that, in his point of view, was justified.

Grendel shows how a change in perspective can radically alter a story. Given the strong attachment humans have to their beloved dogs, it is natural to be touched by Vest's "Eulogy to the Dog," to feel empathy for Charles Burden, and to not care—or not even be aware of—the plight of the other party in the lawsuit. In the America of our time, when dog owners are abundant and sheep scarce, this is the point of view most likely to prevail. But if the research path into Vest's courtroom speech winds through the anguish of nineteenth-century shepherds losing their charges in dog raids, it is easy to commiserate with Leonidas Hornsby, to feel his pain as he encountered yet another wooly carcass done in by hounds, and to understand why he would fire a gun at stray canines on his property. This is not to compare the dog owner or the sheep owner to either Beowulf or Grendel, and it would be patently unfair to compare the two species of animals involved to the hero or the monster from the epic saga. But it is clear that in the America where George Vest practiced as a country lawyer, in contrast to today, many people considered Hornsby's killed sheep a bigger problem than Burden's slain dog, no matter how fond the owner was of the hound. And the Missouri state legislature soon acted as other states had before—providing relief for sheepmen, not for dog owners.

When Georgia's Judge Powell declared that "higher or unwritten law" made sheep-killing dogs outlaws in his state, he noted that some other states simply condemned such animals through normal statutory law. He named three of them: Massachusetts, California, and Missouri.[56] Yes, Missouri, the state where *Burden v. Hornsby* was adjudicated, where George Graham Vest spoke so eloquently of the noble nature of dogs. In 1877, less than a decade after

Vest's celebrated courtroom speech, Missouri's General Assembly passed "An Act to Provide for the Registering and Licensing of Dogs." The act did not speak to Burden's problem, receiving restitution for the wrongful shooting of a beloved pet. Part of it, however, forcefully addressed Hornsby's complaint, marauding canines killing his sheep. Section 9 of the act declared:

> In every case where sheep or other domestic animals are killed or maimed by dogs, the owner of such animals may recover against the owner or keeper of such dog or dogs the full amount of damages, and the owner shall forthwith kill such dog or dogs; and for every day he shall refuse or neglect to do so [after notice], he shall pay and forfeit the sum of one dollar, and it shall be lawful for any person to kill such dog or dogs.[57]

All those romanticized accounts of Vest's eloquence leave out this significant postscript: that his beautiful words may have swayed a jury, but in 1870s Missouri it was the damage caused by dogs—not the damages due someone who lost a dog—that most concerned the state's elected representatives. If this law had been in effect just eight years earlier, and if Hornsby could satisfy a court that Old Drum had harmed his sheep, Burden, the dog's owner, would have been legally required to shoot the animal if it had remained alive. His best friend. In spite of all the acclaim showered on Vest's speech, regardless of the commemorative plaque and the Old Drum statue proudly displayed in Warrensburg, in a real sense the interests of the sheep farmers prevailed over the affections of the dog owners. Missouri decided that no matter how noble your hound, if it hurt sheep it was a nuisance and had to be destroyed. As we will see, this legal policy persists today.

It is not apparent whether the Missouri legislature was influenced by the verdict in the Old Drum case. But even if any of the representatives were aware of what happened in that Warrensburg courtroom, the bill that was passed mandated that friendships between humans and dogs could not stand in the way of the economic importance of the sheep.

In 1875 Tennessee passed a bill titled "An Act to Increase the Revenue of the State and to Encourage Wool Growing." As we have seen, encouragement of rearing sheep for their wool usually went hand in hand with canine control. This legislation assessed a dollar-per-head tax on each dog owned, but it did not survive a court challenge. The Tennessee Supreme Court declared that it violated the state's constitution, which specifically mandated: "All property shall be taxed according to its value." (A dissenting justice complained that if dogs could only be taxed that way, "it would probably be found impracticable to find any means of ascertaining their value.")[58]

Adjacent to Tennessee, Georgia took no chance that any court could nullify a dog tax. This was done by explicitly granting legislative power to regulate canines in the state's 1877 constitution, drafted just a few months after the Tennessee court ruling. A delegate to the constitutional convention proposed that Georgia's new supreme law include a provision that all dogs be taxed "[not] less than five dollars nor more than ten dollars per head."[59] The law struck down in Tennessee taxed dogs one dollar per head; now here was a neighboring state considering a constitutional provision requiring that dogs be taxed at least five times that fee—an indication of just how seriously some Georgians regarded dogs as a problem. For reasons not clear, the constitution adopted did not use the simple word "dogs," verbosely instead giving the state legislators the authority to "impose a tax upon such domestic animals as, from their nature and habits, are destructive of other property."[60] But clearly dogs were the animals described as "destructive," and sheep were the property expected to benefit.

Since policies instituted in southern states during and after Reconstruction disadvantaged freed African Americans hoping to become self-employed shepherds, the question arises: Was there a racist aspect to dog laws in the postbellum South?

Probably this was so. The delegate to the 1877 Georgia constitutional convention who proposed that five-dollar-minimum dog tax certainly must have known that such a fee would be a particular burden for the tens of thousands of his state's residents who had themselves been legally designated as property just a few years earlier. It has also been shown that during the antebellum years there was a general practice to keep enslaved people from owning dogs at all, the reason for this policy being to keep African Americans from possessing an animal harmful to sheep. When the nature of the relationship of white and black people changed following the Civil War, the nature of the relationship between dogs and sheep remained as it had always been; flocks were still at risk from canine slaughter. No doubt this caused some whites in power to remain hostile toward the notion of African American ownership of dogs; a tax on them would be one way to keep those carnivorous animals out of black hands.

But while there likely was a racist motive for taxing dogs in the South of the 1870s, it is clear that earlier nineteenth-century support for dog taxes did not arise from racism. In the antebellum years, when an enslaver like Thomas Jefferson had direct control over what property he would allow his enslaved, he plainly saw the marauding animals kept by his free white neighbors as the problem needing a legislative solution.

Judge Powell ruled in Georgia in 1909 that the absence of legislation specifically permitting the destruction of a sheep-killing dog was irrelevant—such a dog could be shot as a matter of "higher law." There would be no need for such jurisprudential gymnastics today because both the canine and ovine circumstances of *Miller v. State* are covered by statute. Section 4-8-5 of the Official Code of Georgia declares that no person shall "harm, maim, or kill any dog, nor attempt to do so." To do this would constitute cruelty of the very sort that Mr. Miller's son stood accused of. But the modern law has exceptions, one of which is that a person in Georgia may legally "kill any dog causing injury or damage to any livestock, poultry, or pet animal."[61] While Georgia defines "livestock" as "goats, sheep, mules, horses, hogs, cattle, and other grazing animals," because of their value and their particular vulnerability to dogs, throughout American history it was sheep that were considered to be most in need of protection.[62]

Georgia is not unusual in this regard; most states have similar statutes. Consider New Jersey, a heavily urbanized and densely populated northern state. There the law reads that one may humanely destroy a dog "found chasing, worrying, wounding or destroying any sheep, lamb, poultry or domestic animal."[63] Thus, if a city dweller in Georgia, New Jersey, or other states were to take a beloved pooch on a trip to the country and the dog were to run after sheep, the owner of the flock could shoot the dog dead and be in no legal jeopardy. In the United States of the twenty-first century, where there are so many pet dogs and one can drive rural highways for hundreds of miles without spotting a single sheep, this may seem an anomaly. But it is an existing reminder that, for much of American history, advocacy for increased flocks of sheep went hand in hand with irritation toward—or even hatred for—dogs and their owners.

CHAPTER 6

Mary Had a Little Lamb and the West Had Big Burly Cowboys

Sheep, the Frontier, and Gender Bias

∽∞∞∽

> There are . . . silent sheep-herders, with cast-down faces, never able to forget the absolute solitude and monotony of their dreary lives, nor to rid their minds of the thought of the woolly idiots they pass all their days in tending.
> —THEODORE ROOSEVELT, *Ranch Life and the Hunting Trail*

Occasionally a newborn lamb is rejected by its mother. Or the ewe doesn't produce sufficient milk to sustain her baby. Sometimes in this situation, another ewe can be convinced to foster the newborn. Otherwise the lamb will die. That is, it will perish unless humans intervene and raise the animal by hand. With sheep such an important part of early nineteenth-century America, inevitably many shepherds and their families became foster parents for an orphaned lamb. The young children in the family must have been particularly delighted with this pet—or this gentle animal that remained a pet until it got older. Since lambs bond with their caretakers, no doubt there were tame little lambs that followed their special young human friends. One of these children may well have been named Mary. After all, this was an extremely common name for a girl in eighteenth- and nineteenth-century America, especially in New England, with its Puritan roots.[1]

The stage was thus set for a lamb to follow a little girl named Mary as she strolled to school. And that led to what has been called the most famous children's poem in the English language.[2] The British writer and humorist E. V. Lucas went further; he concluded that "Mary's Lamb"—often identified by its first line, "Mary Had a Little Lamb"—was the best-known four-line verse in English.[3]

One morning around 1810, a young schoolteacher in New Hampshire was surprised when one of her students—a girl named Mary—walked into the one-room schoolhouse followed closely by her pet lamb. Cute as this might have been, the teacher realized this would distract the children from their lessons, so she insisted the lamb be left outside. After class, Mary walked out of the small building—and promptly the lamb, which had been patiently waiting, ran up to its favorite little girl. The other children were astonished and asked their teacher why the lamb loved Mary. The teacher, realizing a perfect opportunity to give the children a moral lesson, told her class that the lamb was simply reciprocating the fondness shown to it by its human friend. Twenty years later that teacher, Sarah Josepha Hale, published her famous poem based on the incident.

This, at least, is one published account on the origin of the poem.[4] But there is another. When Hale was elderly, one of her children responded to an inquiry by giving a different version. In this retelling, while the verse was based on personal experience, Hale herself was the Mary, the little girl whose lamb followed her to school.[5]

Whether she was actually the teacher, the little girl, or simply the poet, Hale achieved renown for more than just a beloved nursery rhyme. Born Sarah Josepha Buell in Newport, New Hampshire, in 1788, her early education was through what today would be called home schooling. As a little girl, Sarah was reading the Bible, Shakespeare, Milton, Pope, and *Pilgrim's Progress*. One thing troubled her about the classics, she recalled later: "Few were written by Americans, and none by *women* (her emphasis)" She thus adopted a twin goal: "to promote the reputation of my own sex, and do something for my own country."[6]

Her strong literary background helped immensely in 1822 when her husband, David Hale, died suddenly of a stroke. Mr. Hale had been a reasonably successful attorney, providing well for his family in his lifetime, but he didn't leave much of an estate. With him gone, Sarah, now thirty-four, had to sustain not just herself but also the five Hale children, ranging from an infant to a seven-year-old. What was a woman of her time to do? She acknowledged, "[There are] very few employments in which females can engage with any hope of profit." But perhaps all that reading and studying she had done her whole life would bear fruit. "My own constitution and pursuits, made literature appear my best resource," she determined. In 1823 she published her first book of poems, and she followed this four years later with a two-volume novel.

Hale's efforts paid off in 1828 when she was invited to move to Boston to be the editor-in-chief of the *American Ladies' Magazine*—the first magazine for women, edited by a woman. The periodical featured essays, fiction, and fashion illustrations in color. Almost a decade after Hale took charge, it merged with *Lady's Book*, a periodical owned by the Philadelphia publisher Louis Godey. Godey named the new combined journal *Godey's Lady's Book* and invited Hale to stay on as the literary editor. One chronicle of Hale's career proclaims: "Godey's would become the most widely-read magazine of the 19th century and Sarah one of America's most influential voices."[7] Today, her influential voice is best remembered for two things. First is her advocacy—and ultimately her success—in promoting a national day of Thanksgiving every November.[8] Second is her authorship in 1830 of *Poems for Our Children*, a volume of children's poetry that included "Mary's Lamb."[9]

Or did she write it? Some argued that without giving credit, Hale lifted much of it from another source. Among the doubters of her authorship was an American industrialist who in the 1920s was called the most famous man in the world.

Become wealthy from your business, and you will have plenty of time to devote to your other passions. While Henry Ford's automobile company made him his fortune, that wealth enabled him to indulge his interest in America's past, especially the nineteenth century of his childhood. Ford took up the hobby of collecting old editions of *McGuffey Readers*, a popular series of nineteenth-century textbooks, amassing probably America's largest collection of the volumes. Ford also spearheaded the restoration of his own boyhood home. This, like collecting old schoolbooks, enabled him to connect with an America that was fading away—its fast disappearance, ironically, due in large part to his and other industrialists' success.[10] In the 1920s Ford's enthusiasm for the recent past was ratcheted up a notch when he saved a historic inn associated with one of the most famous Americans of Ford's youth.

Henry Wadsworth Longfellow was already the most well-known American writer when he published *Tales of a Wayside Inn* in 1863, the same year Ford was born.[11] The collection contains some of Longfellow's most familiar poems, including "Paul Revere's Ride," that romanticized introduction to the American Revolution that many of us read in grade school.[12]

The inn referenced in the title of the book—the Wayside Inn of Sudbury, Massachusetts, was not imaginary; it was (and still is) a real place. By the

1920s, this historic structure had fallen into disrepair. Leaders of a local preservation trust, knowing Henry Ford's interest in history, asked if he would purchase and restore the inn. And so he did.[13]

In 1923 Ford bought the building and ninety acres adjacent to it for $65,000. Disturbed by the presence of some seedy businesses nearby, he eliminated them by eventually acquiring twenty-five hundred more acres. Altogether Ford wound up spending over $2 million on the Wayside Inn project (the equivalent of $36 million in 2023).[14]

It occurred to Ford that he could purchase other old buildings and move them to the Wayside Inn site, thereby creating a historic village. Thus, when he learned that the Redstone Schoolhouse of Sterling, Massachusetts, was available, he bought it and had it moved to a spot a short walk from the inn in Sudbury. This action had an odd consequence because the Redstone Schoolhouse was the site of an alternative origin story for "Mary's Lamb."[15]

In 1878, nearly fifty years after the publication of *Poems for Our Children* and with the elderly Sarah Hale just a year away from her death, Mary Sawyer Tyler, a woman in her seventies, asserted that she was the Mary of the poem, the little girl whose lamb followed her to school.[16] Tyler insisted that this had happened around 1817 and that among those witnessing the fondness her lamb had for her was John Roulstone, a boy of about twelve. The very next day, Tyler recalled, Roulstone gave her a little gift—a slip of paper on which he had written the first three stanzas of "Mary's Lamb." This was well over a decade before the verse appeared in print.

Ruth Finley, in her 1931 biography of Hale, was not pleased with this doubt cast on the authorship of the famous nursery rhyme. "Mrs. Tyler never produced the slip of paper or any witness who had ever seen it," she retorted.[17] Indeed, it does seem a bit odd that Tyler's tale would be taken seriously without physical evidence—especially since she was claiming that a large portion of a poem attributed to a grown woman in her forties was actually the work of a preteen boy. Controversy, however, often piques interest even when plausible verification is lacking, and that is what happened here. In defense of her account, Mary Tyler gave bits of wool from her childhood lamb—which by then had long been dead—to children and families, telling them all the story of her pet who followed her to school. Naturally the wool could have been from any sheep. Not unreasonably, Richard Hale, Sarah Hale's grandnephew, mocked the logic that this proved anything about the authorship of the poem.[18]

Richard Hale wrote this defense of his great-aunt in 1904, over a quarter-century after Mary Tyler came forward with her story, demonstrating that this was not just a fleeting literary controversy. Nor was it about to end. Twelve years later, *Poems for Our Children* was reissued, and again Richard Hale fought for his family's honor, this time referencing his earlier essay in a preface ostensibly addressed to young readers. "Dear Children," he began, "in 1904 I wrote for your elders the story of how my great-aunt wrote her poems and printed her book, and also the further story how some other people pretended that someone else wrote 'Mary Had a Little Lamb.'"[19] Mr. Hale was determined not to let another generation grow up with doubts about who wrote the poem.

All of this probably would not have concerned Henry Ford, were it not that the little schoolhouse Mary Tyler attended, followed there one day by her wooly pet, was none other than the Redstone Schoolhouse Ford had moved to his Sudbury site. He now had a vested interest in promoting this as the setting of "Mary's Lamb."

This Ford did in two ways. First, he had a plaque installed on the schoolhouse acclaiming it as the birthplace of the nursery rhyme. Second, in 1928 he published a forty-page booklet, available for sale at the Wayside Inn, in which Ford's anonymous hired writer made as compelling a case as possible that Mary Tyler's story was gospel.

The title page of Ford's booklet, *The Story of Mary and Her Little Lamb*, advertises that it includes "a critical analysis of the poem." As a nursery rhyme, "Mary's Lamb" does not normally receive scholarly attention, and it does not get much here. Mostly the "critical analysis" is an effort to convince the reader that young John Roulstone is the rightful author of "Mary's Lamb," or at least those first three stanzas, the best-known portion of the work.

Ford's ghostwriter's effort is a bit maudlin and not particularly persuasive. "If one were asked the reason for belief in the story as here told, the first answer would be, *Mary Sawyer said so*," the booklet pleads (emphasis in original). Furthermore, the writer adds as if relevant, "Mrs. Tyler lived a most highly respected lady and died at the age of eighty-three in December, 1889. No one who knew her character could doubt her truthfulness."[20] One can picture Richard Hale reading this and sputtering, "Well, my great-aunt lived a most highly respected lady and died at the age of ninety in April 1879, no one who knew her character doubted her truthfulness either, and, for crying out loud, *she wrote 'Mary's Lamb'!*"

In response to the argument that Henry Ford questioned Sarah Hale's authorship of "Mary's Lamb" because he had a strong motivation to promote a different origin, one could, of course, reply that Richard Hale had an equally strong bias to defend the honor of his deceased great-aunt. It could further be contended that Ruth Finley, as Sarah Hale's biographer, also had reason to uphold the veracity of the subject of her book. But it is hard to challenge Finley's tart criticism of the Ford booklet's assertion that Sarah Hale couldn't prove her authorship. "Ordinarily the signature of a professional writer of recognized integrity is sufficient testimony of authorship, and must stand unless conclusive evidence is adduced to the contrary," Finley emphatically wrote.[21]

Finley also maintained that doubts about the authorship of "Mary's Lamb" resulted in part from the inadequate copyright laws and loose publication practices of the era. By 1834 uncredited use of the nursery rhyme had already appeared, and this was just four years after the poem was first published.[22] In 1927, while the Mary Tyler controversy simmered, a seventy-five-year-old gentleman visited the offices of the *Boston Daily Globe*, bringing along a copy of one of his old school readers, published in 1851. It contained the poem "Mary's Lamb"—except there it wasn't called that; it was instead titled "Lucy's Lamb" and opened: "Lucy had a little lamb / Its fleece was white as snow."[23] Furthermore, no author was credited. In a time when the copyright code was so rickety that Mary could quietly morph into Lucy, it is easy to see how the origin of the poem could come into question.

Eventually Henry Ford tried to distance himself from the debate he helped inflame.[24] This is unsurprising, given the connection of "Mary's Lamb" to an agonizing moment in his life. Before it was a booklet for sale at the Wayside Inn, the text of *The Story of Mary's Lamb* had been printed in Ford's own magazine, the *Dearborn Independent*, in March 1927. This was the same time that the *Independent*—and Ford himself—came under fire due to the periodical's anti-Semitic diatribes, one of which resulted in Ford facing a libel suit. In response, Ford issued a thorough apology, declaring that he was not a foe of Jewish people and was guilty only in not exercising diligent editorial control over the writings of contributors to his magazine.[25] Obviously Ford at that time had a more pressing and potentially business-damaging matter on his mind than the authorship of a nursery rhyme.

Coincidentally, Ford's 1927 apology refers to sheep. After declaring that he held no demographic group of people in contempt, he wrote: "Of course there are black sheep in every flock as there are among men of all races, creeds, and nationalities who are at times evildoers. It is wrong, however, to judge a people by a few individuals."[26] By then in his mid-sixties, caught in a nasty storm, perhaps Henry Ford couldn't help but pine for the simpler time of his youth, exemplified by an adorable little lamb trotting behind an innocent girl on her way to school.

Some modern biographies of Ford do not mention his connection to Mary and her lamb.[27] It is obviously just a footnote in the voluminous story of an immensely influential American. But where American sheep are concerned, the affair highlights transformational significance. "Mary's Lamb" was published just four years after the death of Thomas Jefferson, that tireless advocate for the manly patriotism of sheep rearing. The little lamb that followed the girl he loved seemed to herald the emergence of a different view of sheep in popular culture—a view that saw sheep as a rather effeminate animal, one best suited to the care of Mary and other women and girls. In the young America of the founding generation, sheep had been prominently associated with the masculine world of Jefferson's nation-building. Going forward, some prestigious people would instead see cattle culture as the epitome of United States manhood and sheep as a concern only for those who were not male—or at least not manly. No one better demonstrated the "macho cattlemen" and "meek sheepmen" turn in the American psyche than the gentleman who moved into the White House ninety-two years after Jefferson vacated it.

It is startling that Theodore Roosevelt, born into a wealthy New York City family, would become so readily associated with the American West. But in 1883 the future president first set foot in the Dakota Territory, took a liking to the world of the cattlemen, and wound up establishing his own ranch there. For the next several years, Roosevelt divided his time between his political career in New York and his livestock concerns in the Great Plains.[28]

Already a published author, Roosevelt readily secured agreements to write books about his western experiences. This led to a series called "the Badlands trilogy," consisting of *Hunting Trips of a Ranchman* (1885), *Ranch Life and the Hunting Trail* (1888), and *The Wilderness Hunter* (1893). In these lively volumes, Roosevelt vividly chronicled the hard, dusty, horseback life on the plains in the late nineteenth century. The books are a homage to cowboys and cattlemen, along with the author's detailed descriptions of hunting trips to bag everything from bison to grizzly bears.[29]

Roosevelt also used the opportunity to tell his readers how much he hated sheep and the people who took care of them. He wrote: "The sheepherders are a morose, melancholy set of men, generally afoot, and with no companionship except that of the bleating idiots they are hired to guard. No man can associate with sheep and retain his self-respect. Intellectually a sheep is about on the lowest level of the brute creation; why the early Christians admired it, whether young or old, is to a good cattle-man always a profound mystery."[30]

The early Christians admired sheep, of course, because they were a primary source of clothing and food. Sheep were a cornerstone of their economy, just as Roosevelt's cattle were an important part of his balance sheet. Furthermore, as we have seen, sheep were a significant part of the business and commerce of the nation Roosevelt wrote about, represented, and fought for.

It would even be accurate to say that Roosevelt lived in precisely the time when sheep were most essential—an era before the development of synthetic fibers but well after the rising human population had decimated wild fur-bearing mammals as a supply for warm clothing. Indeed, Roosevelt himself acknowledged the value of sheep—indirectly—when he wrote in *Hunting Trips of a Ranchman* about preparations for a winter deer hunt. "The weather was bitterly cold," he recalled, "the thermometer sometimes going down at night to 50° below zero and never for over a fortnight getting above minus 10° [Fahrenheit]." To cope with such a frigid environment, Roosevelt and his companions "wore the heaviest kind of all-wool under-clothing."[31]

The wool in Roosevelt's wardrobe was typical for a cowboy of the northern plains. In the 1950s, Don Rickey Jr., park historian for the Little Bighorn Battlefield National Monument, researched the clothing and equipment of North Dakota cowboys in the 1880s. Rickey considered his most significant research tool to be the interviews he conducted with six men he described as "part of the open range era." Those men ranged in age from seventy-eight to ninety-two, so some of them would have been riding the plains at the same time as Roosevelt.[32] Those cowboys spoke of wearing not only wool underwear but also thick wool socks, wool mufflers, and wool-lined gloves—all things needed to bundle up for cold weather. But the men also recalled that, regardless of the season, many of their shirts and most of their trousers were wool. Sheep provided everyday wear for cowboys, not just protection from freezing temperatures.[33]

Roosevelt thus galloped his horse across plains that featured both cattle and sheep. But how many of those livestock then lived in the West? The Census Office in 1880 admitted that to determine this, first a problem with counting methods needed to be addressed.

⁌⁍

A casual glance at the *Statistics of Agriculture* report of the 1880 census suggests that there were a few hundred thousand more cattle than sheep in America at that time. These returns tallied more than 35.9 million cattle and nearly 35.2 million sheep.[34]

But these statistics, their compilers emphasized, were only a count of livestock on *farms*. The figures did not encompass "the grazing of cattle and sheep over extensive ranges of public or private lands, generally the former, upon the extreme frontier of settlement."[35] In other words, the regular census did not count animals on the ranch or on the range. The Census Office deemed this insufficient; it wanted a tally of *all* of the livestock in the United States. And so special agent Clarence W. Gordon was assigned to organize a count of the cattle, sheep, and swine west of Missouri roaming across the unfenced prairies and plains rather than residing on farms. His work, aided by scores of people who helped him gather the data, resulted in a remarkable 156-page bulletin, *Report on Cattle, Sheep, and Swine, Supplementary to Enumeration of Live Stock on Farms in 1880*.[36]

Gordon's task was daunting. He admitted that strict accuracy was impossible, writing that only a "reasonable approximation" could be made. Summing his estimations, Gordon figured that there were slightly more than 3.65 million cattle ranging across the plains, hills, and arid lands from Missouri to the Pacific Ocean. Added to the nine million head on the region's farms, that meant that the western states were home to over 12.6 million cattle. Gordon further reported that the region had nearly seven million "ranch and range" sheep; adding these to nearly 12.3 million farm-dwelling animals meant that the western states had over 19.25 million sheep (see table 6.1).[37]

Thus, in 1880, the era we associate with the old West, a time popularly associated with cowboys, whose very name carries a bovine reference, cattle were not the ruminants most common across America's vast territory beyond Kansas City. Far from it. The land that was home to the cowboys and

TABLE 6.1. Cattle and sheep in the western states and territories in 1880

State or territory	Farm cattle	Ranch and range cattle (estimated)	Total cattle (estimated)	Farm sheep (estimated)	Ranch and range sheep (estimated)	Total sheep (estimated)
Arizona	44,983	90,774	135,757	76,524	390,000	466,524
California	664,307	150,737	815,044	4,152,349	1,575,000	5,727,349
Colorado	346,839	444,653	791,492	746,443	345,000	1,091,443
Dakota	140,815	65,968	206,783	30,244	55,000	85,244
Idaho	84,867	106,290	191,157	27,326	90,000	117,326
Indian Territory	———	487,748	487,748	———	55,000	55,000
Kansas	1,451,057	82,076	1,533,133	499,671	130,000	629,671
Montana	172,387	255,892	428,279	184,277	95,000	279,277
Nebraska	758,550	354,697	1,113,247	199,453	48,000	247,453
Nevada	172,221	44,602	216,823	133,695	97,000	230,695
New Mexico	166,701	181,235	347,936	2,088,831	1,850,000	3,938,831
Oregon	416,242	181,773	598,015	1,083,162	285,000	1,368,162
Public lands*	———	58,450	58,450	———	———	———
Texas	4,084,605	810,093	4,894,698	2,411,633	1,240,000	3,651,633
Utah	95,416	37,239	132,655	233,121	290,000	523,121
Washington	134,554	63,630	198,184	292,883	96,000	388,883
Wyoming	278,073	243,140	521,213	140,225	310,000	450,225
Totals	9,011,617	3,658,997	12,670,614	12,299,837	6,951,000	19,250,837

* The report defines "public lands" as the "region north of the panhandle of Texas, not included in any state or territorial government."

Modified from table in Clarence W. Gordon, *Report on Cattle, Sheep, and Swine, Supplementary to Enumeration of Live Stock on Farms in 1880* (Washington, D.C.: Government Printing Office, 1883), 149.

the open range had over 6.5 million more sheep than cattle—no doubt to Theodore Roosevelt's disgust.

Economics influenced the relative abundance of sheep and cattle in the West. Sheep had the advantage of needing less real estate. For example, Gordon concluded that in Colorado, each head of cattle required forty acres of rangeland; in contrast, each individual sheep required 7.5 acres of comparable range.[38] In other words, the same amount of land that could only sustain one head of cattle could instead provide food for five sheep with room to spare. Mitigating this point in the sheep's favor, in the right circumstances cattle could be more profitable even with their need for additional space. Again in Colorado, Gordon calculated the value of a single cow, depending on its breed, at $12–21. In the same state, a single ewe's value was computed at $2.25–3.[39] Thus, it would take four of the finest ewes to equal the value of just one cow of even the lowest grade.

A more important comparison of these two livestock is that sheep generally are hardier than cattle. While Theodore Roosevelt preferred cattle for the Dakota range, the Dakota range did not share his bias. The 1886–87 winter across the northern plains was brutal, causing 60 to 75 percent of the cattle to perish. Roosevelt himself lost over half his herd.[40] Commenting on Roosevelt's loss, historian Stephen Ambrose wrote that after that dreadful winter "most of the cattle ranchers on the High Plains turned to herding sheep, which were better able to survive blizzards and fetched a high return on investment."[41] There is some truth to this, but the reality is a bit more complex, as shown by a comparison of the number of Montana sheep and cattle recorded in Gordon's report and a decade later in the 1890 census (see table 6.2).

In spite of the severe weather-related losses that cattle ranchers of the northern plains had suffered in the 1880s, Montana boasted over a million more cattle in 1890 than it had ten years earlier. The state thus had no substantial turn away from bovines, as a cursory look at Ambrose's comment might suggest. But entirely in line with his statement, the explosion in the state's sheep population was far more dramatic—from fewer than three hundred thousand in 1880 to over 2.3 million in a single decade. The late nineteenth-century American West was a landscape of livestock, with millions of cattle but even more millions of sheep.

While sheep may be hardy, that hardiness has limits. Gordon noted that between December 1877 and January 1878, extremely persistent and severe

TABLE 6.2. Increase in the number of cattle and sheep in Montana between 1880 and 1890

Year	Number of sheep	Numerical increase	Percent increase	Number of cattle	Numerical increase	Percent increase
1880	279, 277	—	—	428, 279	—	—
1890	2,352,886	2,073,609	742%	1,442,517	1,014,238	237%

winter storms buried the pasturage under snow. As a result, a sheep ranch near Colorado Springs lost five hundred head out of a flock of thirty-seven hundred. Other sheep owners in the state reported even higher losses totaling, about 20 percent of their flocks.[42]

Gordon's report is more than just a dry computation. It is a comprehensive history of the growth of the West as livestock range, showing that those fabled long-distance drives of the era were often of sheep rather than cattle. Consider this passage: "A firm drove about 30,000 sheep from California to Montana in 1878, but, becoming dissatisfied with the country, turned south again in 1880 with 32,000 head and located in Apache county, Arizona, having driven by the Mormon Trail through Utah."[43] Changing the word "sheep" to "cattle" in that paragraph puts it more in line with the popular image of the old West. But in this instance and many others, the "cowboys" were moving sheep.

Note that this drive was eastward—sheep originating in California traveled to Montana and later to Arizona. But before that could happen there had to have been earlier drives bringing sheep from east to west. When California became a state, the sheep present there were predominately from Mexico. To augment the flocks, beginning in 1852, a series of drives brought sheep from New Mexico, Illinois, and Missouri.[44] By the time of the drive Gordon described, sheep were well established in California, to the point that it provided stock for states to its north and east. In 1880 alone, sheep drives sent nearly 150,000 California head to Oregon, Washington, Montana, Wyoming, Colorado, Utah, and Nevada.[45]

Those Illinois and Missouri sheep supplying the 1852 drive to California were themselves the result of much earlier westward movements from the East to the states beyond the Appalachians. Eighteenth-century records

Mary Had a Little Lamb

show that sheep, like other livestock, traveled along with the earliest settlers. Indeed, sheep-hater Theodore Roosevelt wrote in his multivolume series *The Winning of the West* that, in 1788, "967 boats carrying 18,370 souls with 7986 horses, 2372 cows, 1110 sheep, and 646 wagons, went down the Ohio."[46] Americans were on the move, and so were the domestic animals crucial to their livelihood, including sheep.

From Gordon's report it is clear that sheep were far more numerous than cattle in the American West of 1880, but reality is sometimes subordinate to perception. How and why did the popular but inaccurate image of an old West as predominately cattle country originate?

In 1890—the same time Roosevelt was writing about the West—Oscar Wilde published *The Picture of Dorian Gray*, with its memorable line: "There is only one thing in the world worse than being talked about, and that is not being talked about."[47] Were American sheep of the late nineteenth century able to comprehend this, they might agree. After all, they were about to go from being written about harshly by Theodore Roosevelt to being ignored entirely in enormously influential works by two of his close acquaintances. Sheep were soon to be disregarded in one of the most famous essays on American history, and also in a novel that initiated a whole new genre of fiction. This, I believe, has affected how Americans view sheep to this day.

※

Accounts of the reactions of visitors at the World's Columbian Exposition, held in Chicago in 1893, routinely use words such as "dazzled," "amazed," and "awestruck."[48] Certainly the young historian Frederick Jackson Turner, like others, must have been impressed by all the fair's sights. But he was not there just as a visitor. At the exposition observing the four-hundredth anniversary of Columbus's 1492 first voyage to the Americas, Turner gave a talk advancing his view that the westward march of Eurocentric people had hit a wall.

The Columbian Exposition included a series of academic symposia, or, as they were called, congresses.[49] At the "Congress of Historians" Turner delivered a lecture titled "The Significance of the Frontier in American History," later known as "Turner's Frontier Thesis."[50] The Huntington Library in California, custodian of Turner's papers, succinctly describes the thesis as the assertion that "American society owed its distinctive characteristics to experience with an undeveloped frontier."[51] It has been written that within thirty years Turner's thesis was widely accepted as the primary explanation for America's growth; it has also been called the most influential piece of historical writing in American history.[52]

Turner took as an inspiration for his thesis a remark in the 1890 census. "Up to and including 1880 the country had a frontier of settlement," a census report asserted, "but at present the unsettled area has been so broken into by isolated bodies of settlement that there can hardly be said to be a frontier line."[53] This end of an era, Turner argued, was of huge significance. "Up to our own day," he declared, "American history has been in a large degree the history of the colonization of the Great West. The existence of an area of free land, its continuous recession, and the advance of American settlement westward, explain American development."[54]

Sheep, as we have observed, were very much a part of this westward settlement, but Turner did not mention them. On the other hand, his essay contains several references to cattle. Turner's overlooking of sheep is most notable when he discusses changes in Wisconsin, the state of his birth. He writes: "What is now a manufacturing State was in an earlier decade an area of intensive farming. Earlier yet it had been a wheat area, and still earlier the 'range' had attracted the cattle-herder. Thus Wisconsin, now developing manufacture, is a State with varied agricultural interests."[55]

In 1860, the year before Turner was born in Portage, Wisconsin, the state had just over 642,000 cattle and almost 333,000 sheep. Only ten years later, Wisconsin counted nearly 832,000 cattle, less than a 23 percent increase. The state's sheep population, in contrast, had tripled to well over a million animals.[56] Thus, the big change in the state's animal agriculture in the decade of the Civil War—during Turner's boyhood—was that sheep raising became much more significant. He did not specify this as one of those "varied agricultural interests." To be sure, this ovine boom came after Wisconsin had moved past the "range" period he described; thus the omission of sheep from the discussion would not be as conspicuous if Turner had not here and elsewhere in the essay readily associated the word "range" with "cattle-herder," never using "sheep-herder" or even "herdsman" to encompass all livestock.[57] The "Significance of the Frontier in American History" ignores sheep as an integral part of America's western movement while it brings up cattle at several points, whether the bovines needed to be highlighted or not.

In shutting his eyes to the prominent role of sheep in American settlers' westward march, Turner actually was just getting warmed up. Fewer than ten years after his address at the Columbian Exposition, he wrote an article titled "The Middle West," highlighting European American expansion into that region. Once again cattle appear; sheep do not.⁵⁸ Thus, Turner writes that when the Indigenous Americans were subjugated, "the Great Plains were open to the cattle ranchers," that a result of the conquests was prairie turned into "vast cattle ranches," and that the most prominent Midwestern cities stood at "the meeting point of corn and cattle."⁵⁹ Most notably, in commenting on the rise of the meatpacking industry in those cities, Turner writes that these businesses sent "the beef and pork of the region to supply the East and parts of Europe."⁶⁰ The next chapter will describe how the meatpacker also sent the lamb and mutton of the region to supply the East and parts of Europe, but Turner writes nothing of this.

Moreover, the essay mentions essentially every significant aspect of the region's economy *except* sheep. Besides cattle, Turner writes of fields of wheat, corn, and oats; the vast pine forests of the upper Great Lakes providing lumber; the iron deposits around Lake Superior; the gas and oil fields of the Ohio Valley; the coal in Illinois and elsewhere; the lead and zinc of the Ozarks; and even the gold of the Black Hills.⁶¹ Missing is even a passing allusion to the region's copious flocks of wool on the hoof.

Arguably, the omission of sheep in Turner's essays only affected the perceptions of scholars reading his work. Popular culture was more influential than academic articles in affecting how ordinary Americans would come to see the late nineteenth-century West. This is why it is significant that there are also no sheep in the 1902 novel *The Virginian: A Horseman of the Plains*.

⁂

Owen Wister and Theodore Roosevelt had striking biographical similarities. Roosevelt was born in New York City in 1858 and first laid eyes on the West in 1883. Wister was born in Philadelphia in 1860 and made his initial trip to the West in 1885. Both men were thus urban easterners who came to the West in that decade between the Census Office commissioning Clarence Gordon to count range sheep and cattle, and that same government body ten years later announcing that there no longer was a frontier.⁶² Also like Roosevelt, Wister wrote about the West. But unlike Roosevelt, whose books about the region were nonfiction, Wister penned fictional accounts, most famously *The Virginian*.⁶³

While Wister's first foray across the plains took him to a ranch in Wyoming—the kind that would later serve as the setting for his best-known work—he traveled extensively in the region. In 1893, the same year that Turner expounded his Frontier Thesis, Wister traversed from Texas to Wyoming and then to Arizona. The introduction to a 1917 edition of *The Virginian* takes note of these forays, commenting, "Thus he came to know intimately the whole of the ranching country, its landscape and the soldiers, Indians, and cowboys who dwelt in it."[64] This was, again, a landscape where sheep outnumbered cattle. Wister introduced *The Virginian* by writing, "Any narrative which presents faithfully a day and a generation is of necessity historical; and this one presents Wyoming between 1874 and 1890."[65] In 1880, when still a territory, Wyoming had an estimated 520,000 cattle but also 450,000 sheep. In 1890, the year Wyoming was admitted as a state, the number of cattle had increased to 685,000, but the state's population of farm sheep had ballooned to over 712,000.[66] Thus, just as a big change in the Wisconsin of Frederick Turner's youth was an explosion in the state's sheep population, the Wyoming that Wister wrote of had a similar massive increase in sheep numbers, leading them to become more numerous in the state than cattle.

As with Turner's essays, this change goes unnoticed in *The Virginian*. To be sure, the novel is packed with bovine allusions. "Montana is all cattle," the Virginian exclaims at one point. Elsewhere it is noted that "calves had begun to disappear in Cattle Land, and cows had been found killed." Men are said to gather to discuss "the cattle interests."[67] Chapter 4 is entitled "Deep into Cattle Land." It might be said that cattle are one of the characters in Wister's book. Missing from *The Virginian* are any domestic sheep. The only two references to sheep in the novel are of wild bighorns.[68] Wister depicts a Wyoming in particular and a West in general that is exclusively cattle country, without a sheep in sight.

When first published in 1902, *The Virginian* was an immediate best seller, with two hundred thousand copies sold in its first year.[69] It was also enormously influential. One account of its impact declared that it "established the Western as a distinct genre of popular fiction" and said that "it set the pattern for thousands of short stories, novels, motion pictures, and television programs."[70] These were narratives that showed a West full of stock heroes, heroines, villains, horses, and cattle, but sheep, if they appeared at all, were of minor impact and, when present, a nuisance to cattlemen. Scores of examples of this could be cited. A single one will have to suffice, a 1967 episode of *The Big Valley*, a television program following the fictional adventures of the Barkley family, owners of a vast ranch near Stockton, California, in the 1870s.[71]

In "A Flock of Trouble," Nick Barkley, one of the brothers in the clan, unwittingly becomes owner of twenty sheep, handed over to him by a shepherd in payment of a gambling debt. Being a cattleman, Nick is angry and plans to divest himself from this unwanted burden as speedily as possible. Alas, he runs into some difficulties in his preparations to have the sheep slaughtered. Stuck with them until he can rectify these problems, he moans that the flock is difficult to manage. Get a sheepdog, he is advised. "In the middle of cattle country, where am I going to find a sheepdog?" he snarls at this suggestion.

Having sheep—even a flock of fewer than two dozen head—turns Nick into a local pariah. Walking into his regular tavern, the bartender refuses him service, saying he can't serve a sheepman. The cattleman present then jeer at Nick, and a stereotypical western barroom brawl ensues, pitting Nick against men who earlier in the episode were shown to be his friends and associates. The tavern fight ends eventually, but, as the episode proceeds, the conflict continues to escalate, and gunfire is exchanged. In the climactic scene, one of Nick's former friends, now an adversary, exclaims, "This is my valley too, Nick. My family sweated and bled for it just like yours. And I'm not going to stand by and see it trampled to death by those sheep!" Nick turns diplomatic, replying, "We've got millions of acres here—room enough for cattle *and* sheep."

This is a case where facts can readily debunk fiction. In table 6.1, note that California in 1880 had about 815,000 cattle—and over 5.7 million sheep. Indeed, no other western state had a greater gap between the ovine and bovine populations as California did. Moreover, San Joaquin County, the setting for *The Big Valley*, had over 182,000 sheep on farms, dwarfing the county's population of farm cattle, which stood at fewer than 17,000 head.[72]

Armed with this information, we are easily able to spot a slew of historic inaccuracies in "A Flock of Trouble." The episode presents sheep as outliers to the California agricultural economy when they were in fact a mainstay. Nick Barkley's remark about the difficulty of finding a sheepdog "in the middle of cattle country" is a complete non sequitur. It is practically the equivalent of bemoaning at the Westminster Kennel Club that no dogs can be found among the scores of cats. The bartender who shuns Nick, if he knew what was good for business, would be more likely to refuse service to a cattleman than to a sheepman, since sheepmen were more numerous. Finally, Nick's comment that the area has room enough for cattle and sheep—presented as statement of his magnanimousness—is meaningless. Cattlemen like Barkley were the minority in late nineteenth-century California. They were not in the bargaining position of strength that Nick's assertion suggests.

At least in one aspect, however, "A Flock of Trouble" is accurate: the animosity between Nick and the other cattlemen turns violent. This leads to a gunfight scene, although no one is killed. Battles between cattlemen and sheepmen were a real-life part of the nineteenth century, and the combatants were not always as fortunate as those in *The Big Valley*. The most infamous conflict, occurring in Tonto Basin, Arizona, was the Pleasant Valley War, sometimes called the Graham-Tewksbury feud. Trouble began when the Tewksburys ranged sheep on land claimed by the cattle-ranching Grahams. Once violence began, it continued for five years, and over thirty people were killed.[73]

Notably, while sheep are absent from *The Virginian*, they are mentioned in another iconic novel published just four years later, one set in Chicago rather than in the West. Upton Sinclair's *The Jungle* describes the hard lives of immigrants employed in the city's meatpacking plants. Sinclair writes that every day there were slaughtered "ten thousand head of cattle . . . and as many hogs, and half as many sheep—which meant eight or ten million live creatures turned into food every year."[74] *The Jungle* was a work of fiction, but this passage was based on fact; thousands of sheep actually were slaughtered each weekday at Chicago's Union Stockyards. And those sheep came from the West of Frederick Jackson Turner and Owen Wister, a West with many more sheep than cattle, which one would never know from Turner's essays or Wister's book.

The triumvirate of Roosevelt, Turner, and Wister moved in the same circles. Several months before Turner delivered his seminal address at the Chicago World's Fair, Roosevelt lectured at a meeting of the State Historical Society of Wisconsin. Titled "The Northwest in the Nation," Roosevelt's address presaged themes that Turner would discuss in his more famous speech.[75] "Frontiersmen lived a life which is now fast vanishing away; there is no longer any frontier," Roosevelt pined.[76] But he was delighted to say that the frontiersmen still had "analogues . . . in the farther west. There they are the heroes of rope and revolver . . . guarding the innumerable herds of branded cattle and shaggy horses."[77] Again, no sheep.

Frederick Jackson Turner was in the audience taking notes as Roosevelt spoke, and it has been argued that Turner's Frontier Thesis owed a huge debt to Roosevelt's Wisconsin address.[78] In any case, the two men began corresponding. Historian Douglas Brinkley writes that between them "an alliance was formed in promoting the frontier hypothesis, with T.R. as the popular oracle and Professor Turner influencing fellow academics."[79]

Thanks to *The Virginian*, Owen Wister was just as much of a "popular oracle" in disseminating ideas about the West as Roosevelt, and Wister was also in Roosevelt's sphere. The two had been friends in college at Harvard and remained close; Wister even asked Roosevelt to read and comment on an early draft of *The Virginian*. By this time Roosevelt was president, with all the duties inherent in the post; nevertheless, he read the manuscript and recommended changes, which Wister incorporated. Grateful, Wister dedicated the novel to his friend, now president of the United States.[80]

Focusing on the nexus between Roosevelt, Turner, and Wister is not to suggest that there was a conspiracy among the men to denigrate or ignore sheep. But, clearly, all three of these chroniclers of the old West had similarities in their backgrounds and experiences that led them to have parallel perspectives, and these views included a common dislike of (or lack of interest in) sheep, no matter how significant they were in the West they all celebrated.

※

Why did Turner and Wister ignore the importance of sheep in the West? They could not plead ignorance. Turner quotes a census report in his famous essay on the frontier, and that same Census Office issued the statistics showing that the West had millions more sheep than cattle. There is no reason to think that Turner would not have seen those documents. Something else must have been in play.

Turner, Wister, and Roosevelt all emphasized the masculine characteristics of life in the West. For instance, in "The Significance of the Frontier," Turner writes of "the importance of the frontier . . . as a military training school, keeping alive the power of resistance to aggression, and developing the stalwart and rugged qualities of the frontiersman."[81] For his part, Wister writes in *The Virginian*, "[A] slow cow-puncher unfolded his notions of masculine courage and modesty."[82]

Similarly, Theodore Roosevelt showed a particular fondness for the word "manly." In *The Wilderness Hunter*, for example, he writes, "Mountaineering is among the manliest of sports."[83] Usually, however, he uses this adjective to describe hunting. In one place he derides the white-tailed deer as a game animal, asserting that, because "it is an inveterate skulker, and fond of the thickest cover," to kill one requires "stealth and stratagem" rather than "fair, manly hunting."[84] In contrast, the bighorn sheep, wild cousin of the domestic livestock Roosevelt hated, earned high praise. "The chase of no other kind of American big game ranks higher, or more thoroughly tests the manliest qualities of the hunter," he asserted.[85] The future president couldn't help but analogize the concept of the manliness of the hunt with the state of the world's nations. In the preface of *The Wilderness Hunter*, he declares, "The chase is among the best of all national pastimes; it cultivates that vigorous manliness for the lack of which in a nation, as in an individual, the possession of no other qualities can possibly atone."[86]

Taking all this into consideration, I suggest that the emphasis on cattle and the ignoring of sheep was a by-product of the desire of the West's biggest boosters to stress the masculine aspects of the region as much as possible.[87] In other words, the popular but inaccurate perception of a late nineteenth-century American West with more cattle than sheep represents a subtle gender bias. Cattle are bigger and stronger than sheep. The roster of people—traditionally men—interacting with large bovines includes not only the rugged, dusty cowboy but also the rodeo bull rider, the matador waving his cape, and the adventure seekers running with the bulls in Pamplona, Spain. Although some rams are pugnacious, they are smaller and less formidable-looking than bovine bulls. Male sheep are not used to test the mettle of the rodeo star or the matador. There is no running of the rams (or of the ewes for that matter). Sheep are more likely to be viewed as the gentle creatures of nursery rhymes, eluding Bo Peep or accompanying Mary to school.

Roosevelt and Wister celebrated the nineteenth-century cattlemen, but those cattlemen themselves contributed to this glorifying of their masculinity over the sheepmen. Here the role of yet another livestock species comes into play—the horse. In 1955, with the popular culture influence of *The Virginian* in full force, professors Joe B. Frantz and Julian Ernest Choate Jr. published *The American Cowboy: The Myth and the Reality*, an effort to separate fact from fiction on the range. Frantz and Choate note that the cattlemen took great pride in their horsemanship. The sheep herders, they also note, did not ride horses, instead walking or riding a donkey. This is followed by a very telling comment, an effort to summarize the feelings of the mounted cattlemen toward the shepherds on foot. "Were such creatures men!" the authors ask rhetorically, notably using an exclamation mark instead of a question mark.[88] Yes, they were men, but they were associated with sheep, which had gradually come to be seen as a less manly type of livestock than cattle.

Perceptions of the West as a man's world extended beyond the sheep-versus-cattle dichotomy. Historian Glenda Riley writes that when Turner delivered his address at the 1893 World's Fair "he clearly was talking about a male frontier" and ignored the role of women.[89] Although Riley does not elaborate on livestock, it is notable that her book mentions two women shepherds and also one female pioneer participating in a cattle drive, though the latter was employed as the cook.[90] The contrast is striking—the female shepherds provided direct care for sheep, but the woman connected with cattle did not have a hands-on role with the animals, instead serving the traditional female role of food preparation. Perhaps inadvertently, Riley touches on the livestock gender bias outlined here. She calls the West a "cattle frontier," when she could have more accurately written of it as a "sheep frontier."[91] So pervasive, it seems, is the popular conception of the West as a land of mostly cattle that it even permeates a volume highlighting the long neglect of women in studies of the region—truly ironic if this theory that gender bias affects our perception of sheep is sound.

As evident in earlier chapters, the men of America's founding era would not have understood or predicted this feminization of sheep. George Washington, Thomas Jefferson, Robert Livingston, and others heralded the sheep as a bulwark of nation-building, which they saw as the duty of all good Americans (or, as they surely assumed, all good American men). Why the change in the view of sheep between the generation of Jefferson and that of Theodore Roosevelt?

I suggest that the simplest explanation is that as the United States grew more urban, the number of citizens having hands-on experience with sheep diminished, leaving people more subject to the secondhand perceptions they would get from reading Roosevelt, Turner, or Wister. As I noted in the introduction, since sheep provided nonperishable wool as their primary good, they never needed to be kept in population centers. City dwellers would thus be especially likely to form opinions of sheep secondhand from published articles or books rather than through direct contact with the animals. In this regard, it is instructive to note that the 1910 census, the first one conducted after *The Virginian* appeared, reported that Theodore Roosevelt's New York City had over 128,000 horses and over 6,300 cattle—but only 393 sheep.[92] Philadelphia, Owen Wister's hometown, reported over 50,000 horses, over 5,100 cattle, but a mere 511 sheep.[93] Thus, well into the twentieth century, American city residents still regularly heard the neighs of the horses and the lows of the cattle, but the baa of an ovine became ever more uncommon to them. Small wonder, then, that Wister's portrayal of a cattle-filled West rather than a sheep-filled West seemed reasonable; it matched what city residents saw of livestock in their own urban neighborhoods. But there was a blip in this trend on the horizon.

A 1917 poster features a photograph of a flock of sheep being driven down Michigan Avenue in Chicago (see fig 6.1.). World War I was raging, and the broadside, issued by the War Department, attempted to inspire civilians to raise more sheep to make up for wool shortages caused by the massive conflict. The caption approvingly notes that in Chicago a "Sheep Club" had been organized to encourage not only rural farmers but also even city landowners with reasonably sized lawns to stock their property with sheep. Five young women flank the flock on one side, with two others strolling behind. The caption calls these women "a parade of shepherdesses." Perhaps they are driving the flock, although, since there is also a man on horseback bringing up the rear, and several horse-drawn carts and automobiles pushing forward behind him, it seems more likely that the sheep forge ahead simply to stay clear of the madding crowd.[94]

The women's attire is outlandish. They sport the stereotyped Little Bo Peep look: immaculately clean, fluffy, light-colored dresses with frills, fancy broad-brimmed hats, heeled shoes, and—of course—shepherd's hooks. Mary and her lamb would look right at home among them. The poster's caption declares that this parade did perhaps as much as anything "to stimulate interest in the new field." So that's why these young women were there on Michigan Avenue. The campaign was using pretty women to hammer home the message, a popular marketing technique before and since.

FIG. 6.1. "Shepherdesses" herd sheep down Michigan Avenue, Chicago, in 1917. National Archives, identifier number 31481135.

But there is something more subtle going on in the poster. A 1917 American looking at it does not read of shepherds but rather of "shepherdesses." Male politicians and administrators in Washington would still fret over imported wool and debate whether or not to impose, raise, or reduce tariffs, and sheepmen out West would still sweat on a dusty range as they managed their flocks, but the popular image of shepherding in America had dramatically changed since the nation's founding, especially the image in the minds of city dwellers, like those in Chicago, a city that not so many years earlier had been regarded as part of the rugged West. Care of sheep was, to the masses, no longer seen as the province of the male patriot or frontiersman or (as in some cases before the Civil War) enslaved men. Sheep had become the feminine livestock, the animal of the nursery rhymes that mothers read to their children. Inaccurately, shepherding was now commonly seen not as the challenging responsibility of crusty herdsmen but as the light duty of Bo Peeps. Considering that the 1917 emphasis on the importance of America's sheep occurred because wool was needed for the uniforms of soldiers, and that warfare is traditionally the most masculine of human endeavors, this popular perception was not simply incorrect but also profoundly ironic.

CHAPTER 7

Machines in the Pasture

Sheep and the Industrial Age

⌒୬୬⑨୬⌒

> I think of our herds spread out on the other side of the hills, and strangely I do not picture the fleece on the backs of our sheep as wool. I see more and more spindles turning; more and more yards of cloth and yarn fall smoothly off the loom to be transformed into suits and dresses, into socks and blankets and underwear that will help keep my countrymen warm.
> —HUGHIE CALL, *Golden Fleece*

At some living history museums and zoos, sheep are sheared in front of the public (see fig. 7.1). If you visit on a day this is being done, the shearer, or an interpreter standing nearby, might hold up a pair of manual shears, devices that resemble oversized scissors, or if you prefer, undersized versions of a gardener's hedge clippers (see fig. 7.2). The interpreter will say that these hand tools were used for centuries by necessity and that some shearers still employ them by choice.

"In modern times," the interpreter perhaps then adds, "we have the option of using power shears." Whereupon the visitors are shown a mechanical device that somewhat resembles the gadget barbers use to closely cut human hair, except it is considerably larger (see fig. 7.3). At the tip is a blade housed between combs, the body of the instrument encases a small motor, and the whole thing is designed to be easily gripped. At its tail, the tool is connected to an electrical cord. The shearer plugs the cord into a nearby outlet or into an extension cord, the switch is clicked from on, and the buzzing noise is evidence that the blade is vibrating. A sheep is held, the shears are put to its wool, and just a few minutes later its fleece has been skillfully removed (see fig. 7.4).

FIG. 7.1. Sheep at the Atlanta History Center awaiting shearing. Author's photo.

FIG. 7.2. A very old pair of manual sheep shears. Author's photo.

FIG. 7.3. A modern electric shearing device. Author's photo.

FIG. 7.4. Shearing of sheep the modern way. Author's photo.

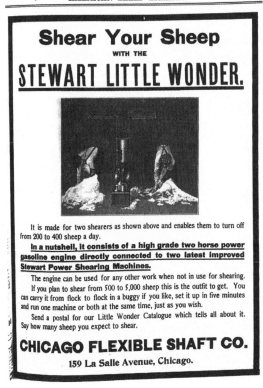

FIG. 7.5. Early shearing machines had to be self-contained, requiring a power source to operate the clippers. This ad from the December 1907 *American Sheep Breeder and Wool Grower* for the portable "Stewart Little Wonder" shows that its gasoline engine accommodates two shearers at once. In 1946 the Chicago Flexible Shaft Company became the Sunbeam Corporation and focused on manufacture of small electrical appliances such as irons, coffeemakers, and toasters.

All good. But look again at that part of this modern contraption that we tend to ignore because it is so familiar to us from devices in our everyday lives—the electrical cord. Powered mechanical shears first appeared in the late nineteenth century, long before electricity was readily available in rural areas that were home to most of country's sheep.[1] As such, there is a gap of well over half a century between those first powered shears and the electric models of our time. Those small handheld clippers that today we think of as powered shears are but a remnant of what was called a "sheep shearing machine" before the mid-twentieth century. The original mechanical shearers were self-contained contraptions with an engine, gears, and belts operating in tandem to provide power to the handheld clippers. Early shearing machines were powered by steam; later models ran on gasoline (see figs. 7.5 and 7.6).

Machines in the Pasture

FIG. 7.6. The challenge of developing a safe, reliable engine to power the movement of clippers through wool was similar to the challenge automakers faced in developing a safe, reliable engine to power the forward movement of wheels across terrain. It is thus unsurprising that some of the same companies that built shearing machines also built automobiles. Here these two are combined: a man is shearing an angora goat using shears powered by a car. Library of Congress, LC-USF34-035589-D.

The story of shearing machines begins in Australia. Frederick York Wolseley, born in Ireland, immigrated there in 1854. He was seventeen years old, and like many other young men in Australia he went into the sheep business. He considered hand shearing an economic liability, reasoning that mechanization could speed up the process, providing more clipped wool in a shorter time. Accordingly, he worked to develop such a machine. Others were also at work designing and building shearing contraptions, but often these prototypes were inflexible and unreliable. Part of the problem, as Wolseley saw it, was that Australia at that time lacked the infrastructure necessary to develop and economically manufacture a shearing machine. To gain access to this, he traveled to England where he had the resources needed to complete his work developing a universal joint that allowed a shearer to move around the sheep being fleeced, as well as a dependable steam engine to power the device. Satisfied with his creation, in 1887 Wolseley founded the Wolseley Sheep Shearing Machine Company to sell it.[2]

Innovators frequently face a daunting challenge convincing people to change the way they have been doing something for years. Proponents of mechanical shearing would be no exception. In Australia, where Wolseley first tinkered, a few sheep stations had shearing machines installed in 1890, but most shearing continued to be by hand.[3] For decades the story would be no different in the United States. Shearing machines were imported from Australia by 1908, but they weren't universally adopted.[4] In 1929, South Dakota shepherd Archer Gilfillan described shearers as "hand men and machine men," suggesting that he encountered both types with equal frequency. There was a prominent difference between the shearing machines Gilfillan encountered and Wolseley's first product—the 1920s devices used engines powered by petroleum, not steam.[5] Over a decade later, while World War II raged with its machines of destruction, rancher Hughie Call wrote, "The veteran woolgrower of Montana still prefers the blade"—that is, they sheared by hand. Her use of the adjective "veteran" is telling. Call noted a generational difference, with younger shearers gravitating toward machines and away from hand clippers. She admitted that the machine shearers worked faster than the blade men, but she bemoaned the change, declaring, "I should hate to see the old, intimate, colorful customs give way to the impersonal efficiency of modern-time machines." Call then confidently asserted that as long as her husband ran their ranch, sheep there would always be shorn by hand.[6]

If it is faster to shear a sheep by machine, and since elsewhere in Call's book she highlights the harsh economics of sheep raising, one might wonder why her husband wouldn't welcome an innovation that sped up the process. Wouldn't time be money for him? Not necessarily—because the practice was not to pay shearers based on their time but rather per sheep shorn. So, to a large extent, the shearer—not the sheep owner—would bear the economic brunt if one's shearing was slower than average.

By discounting a cost-benefit analysis, proponents of hand shearing were free to take a different approach in advocating their old-fashioned way of doing things—namely, that their method was better for sheep welfare. According to the hand shearers Hughie Call spoke with, sheep fleeced by machines were not really sheared; they were shaved—meaning that while hand shearing left a half inch of wool on a sheep's back, a machine removed all the wool down to its skin. The hand shearers asserted that the bit of wool they left on the sheep protected it from the sun, while machine-sheared sheep developed blisters from lack of that protection. Winter comes early in Montana, and the hand men further contended that when the cold weather arrived, machine-sheared sheep had not grown enough wool to protect them from the elements, causing the animals to suffer from exposure.[7] On the other hand, one modern source argues that it was soon apparent that machine shearing caused *less* damage to the sheep, the uniformity of mechanical cutting insuring that the animals were less likely to be gouged by a careless hand.[8] Even before mechanization, when all shearing was done manually, some complained that the general state of the art was poor. "The shearing of sheep, as generally conducted in the United States, *is utterly disgraceful*," said a popular husbandry manual published in 1863. The book protested the abundance of sheep looking as if their fleeces had been "*gnawed off* rather than shorn (emphasis in original)."[9] Taking all perspectives into account, it seems as though the skill of the shearer matters more than whether that shearer uses manual tools or power ones.[10]

Whether or not shearing was a hardship for the sheep, it put significant stress on the muscles of the humans doing it, regardless of whether they cut by hand or by machine. Manual clippers put a constant physical force on hand and wrist muscles, causing repetitive strain injuries. Machine shearing helped alleviate this but led to a new complication—the shearers were more likely to develop back problems, since a mechanically sheared sheep needed to be held in a more rigid position than a hand-clipped one. Also, with the early machines there was a safety issue— if clippers slipped out of a shearer's grip, they would writhe at the end of their hose attachment, putting any flesh in its path, human or ovine, at risk of being sliced, protective combs not yet having been developed.[11]

Both manual and mechanical shearing are grueling, and descriptions of the process routinely emphasize the strenuous, sweaty labor needed to fleece a sheep.[12] Even today, top-flight shearers expend energy and dehydrate at rates similar to those of marathon runners and competitive cyclists.[13]

Shearing is also itinerant. A specialized shearer is not needed year-round but only at that time the sheep in a given place are to receive their clipping. Thus, in the early twentieth-century American West, shearers spent much of the year traveling cross-country. January would find them in California and Arizona; they would then work their way north through the Rocky Mountain states, reaching Montana by July, with some shearers ending the season midsummer in Canada. To earn a living the rest of the year, some shearers worked on ranches, others traveled to the Southern Hemisphere to shear sheep there, and still others apparently were successful enough at their craft that they could get by for the whole year on money earned during shearing season.[14]

Given the demand for shearing, and the time constraints involved, it is no surprise that labor disputes arose by the early twentieth century. "As shearing is a specialized trade," the U.S. Tariff Commission reported in 1921, "and since the season during which it must be done is comparatively short, these men have been in a position to dictate prices which the flockmasters have considered exorbitant."[15] This was written when troubles during World War I were fresh in mind.

"Huns Forced Back in Sharp Attack," blared the headline on the front page of the *Casper (WY) Daily Tribune* on June 6, 1918. The article detailed how half a world away in France, the German troops at the Battle of Belleau Wood were unsuccessful in their attempt to break through Allied lines. An article in an adjacent column notes that the United States would draft 400,000 men that month. Everywhere on the front page, a Wyoming reader of the *Daily Tribune* would have been be confronted by news about the Great War.[16]

Another article was a local story but one that had a direct connection to the war overseas. The short piece noted that Sherriff Hugh L. Patton had made a call on the town's shearing pens to make certain the shearers were working, not striking.[17] This marked the end of a story the *Casper Daily Tribune* had covered more fully a little over two weeks earlier. On May 21, again on the front page, was a report under the headline: "Strike of Sheep Shearers Will Lead to Federal Prosecution if Wool Clip Is Delayed." The flock owners had appealed to Wyoming governor Frank L. Houx to force striking shearers back to work. The state's shearers were holding out for twenty cents per machine-sheared sheep. The woolgrowers' counteroffer stood firmly at seventeen cents per sheep. The article noted that an average day's work for a shearer was two hundred sheep shorn; at their demanded rate of twenty cents per sheep they would earn forty dollars a day. At seventeen cents per sheep, the shearers would make only thirty-four dollars a day.[18] They were fighting for a 15 percent pay increase.

Governor Houx came down firmly on the side of the woolgrowers, and he asserted that he was doing his duty as a responsible statesman during wartime. His executive order fiercely declared, "The wool on the sheep's back is the property of the federal government, and any party or parties who interfere with shearers in removing the clip ... will do so at his own peril and will be dealt with by the federal government."[19] Those several hundred thousand American men being drafted to fight the Germans, after all, needed to be clothed in proper military uniforms.[20] In delivering his warning, Houx in essence declared that when the United States was at war, a reliable supply of wool was a matter of national security—a position the government had taken in previous American military conflicts, and would continue to be so for several decades more. Houx was not exaggerating about the true owner of the wool. Describing the situation prevailing during World War I, the U.S. Tariff Commission later reported: "From April, 1917, until November, 1918, wool passed gradually under Government control, until, at the end of the period, all phases of the business were strictly regulated."[21]

Even the seventeen cents per shorn sheep offered the shearers in 1918 was a much higher price than they could have negotiated in peacetime, when there was no pressing need for military clothing. In 1926, shearers in Opal, Wyoming, would strike, demanding payment of seventeen cents per shorn sheep—the same wage Wyoming shearers were grudgingly forced to settle for just eight years earlier. They ended their walkout when the sheep ranchers agreed to pay fifteen cents per head. This was an increase from the twelve and a half to fourteen cents per sheep that the shearers had been getting—their wages had plummeted since Germany's defeat.[22]

Labor issues would continue to arise for years. A strike in the spring of 1938 brought West Coast shearing to a standstill. It began in late March when three hundred to four hundred shearers in the San Joaquin Valley of California stopped their work. By this time, the shearers were fully unionized, represented by the Sheep Shearers Union of North America. The strikers demanded a closed shop so that there would be a uniform wage of twelve and a half cents per sheep throughout the state, and also the right to have each bag of wool shipped affixed with a label declaring that it had been sheared by union workers.[23]

The strike spread to adjacent states, and by April 9 it affected fifteen hundred shearers—and more than nine million sheep. Shearing had come to a halt not only in California but also in Washington, Oregon, Utah, and Nevada.[24] Woolgrowers responded by bringing in nonunion labor, escorted to San Joaquin Valley under police protection.[25] This union-busting tactic worked, as the idle shearers' funds were dwindling. On April 21 the president of the California branch of the Sheep Shearers Union admitted that the strike had collapsed, with shearers returning to work even though the strike had not been officially called off.[26]

In the Southwest, sheepmen had another way of keeping shearing costs low—employ shearers from south of the border. The U.S. Tariff Commission reported in 1921 that in New Mexico and parts of Arizona, Mexican shearers were paid as little as ten cents per shorn sheep.[27] This information was presented drily. Such was the tenor of the times that there was no suggestion that paying someone less due to ethnicity was immoral or improper.

⁂

While Frederick Wolseley spent the 1880s working on use of engines to power wool clippers, others, notably Karl Benz and Gottlieb Daimler, were fleshing out the intricacies of using engines to propel carriages. Given the mechanical similarities of these efforts, it is no surprise that within twelve years of its inception, the Wolseley Sheep Shearing Machine Company began to manufacture automobiles. The connection between early sheep-shearing machines and the first generation of horseless carriages is even stronger—Herbert Austin, Wolseley's Australia-based mechanic, also went on to build automobiles, founding the Austin Motor Company in 1905.[28]

Another automaker did not build shearing machines, but he was fascinated by the process of shearing because of the need for wool in the manufacture of his cars. In the mid-1920s, with Henry Ford still riding high from the success of his Model T, the Ford Motor Company produced a six-and-a-half-minute silent film featuring a lot of sheep but no automobiles (see figs. 7.7 and 7.8).[29] The opening scene shows a man in a small corral brandishing hand shears as he clips the wool from a sheep. Standing just outside the corral is another man; like the shearer he wears work clothes, but his are accessorized with a necktie and a hat. He must be a manager: occasionally he points out something to the shearer as though giving instructions. Presently the manager leaves the scene, walking back into it a few seconds later accompanied by a third man. This newcomer wears a full suit with an overcoat, an elegant fedora sits atop his head, and he carries a briefcase. While the shearer keeps shearing, the manager jumps over the fence into the corral and grabs another sheep that has not yet been shorn. He then wrangles it over to the side of the corral for the man with the briefcase to inspect. And this he does, burrowing his fingers into the fleece for a few seconds to determine its quality. The sheep squirms and manages to escape the grip of the manager a couple of times, but no matter—the manager also has a pile of shorn wool that he brings over for a look. This too is given a thorough feel by the briefcase-toting man. Obviously he decides this wool is just right for the purpose he has in mind, for after a short chat with the manager he reaches into his briefcase, pulls out a notebook, and writes out a receipt of sale that he signs and then hands to the manager to get his signature as well. Wool has just been purchased, but for what purpose? To this point, a viewer of the film would have no idea; it looks more like something that would be made by a state agricultural extension service than by a private manufacturing corporation.

The film continues with another scene of a sheep being sheared, but this time the action takes place inside a barn and the shearer uses mechanical shearers. He squirts some oil onto the blade, clicks the tool on, and he is ready to go. How much quicker this shearing goes than the one earlier in the film done by hand! There seems to be a subtle message here: mechanizing the process speeds productivity.

FIG. 7.7. A still from a Ford Motor Company silent film, showing use of hand shears. National Archives, identifier number 91950.

FIG. 7.8. Another still from the same film shows a buyer for Ford examining the fleece. National Archives.

Over five minutes into the film a cloth-processing machine is shown, with a caption superimposed that ties the machine to the sheep. "Weaving cloth is one of the world's oldest industrial arts," the caption reads, "yet today its manufacture plays an important part in modern motor car construction." Ford made the film to explicitly show the link between the Model T and sheep.[30]

Ford wasn't finished educating Americans about the connection between sheep and automobiles. Over a decade after making the silent film, the Ford Company built a pavilion for the 1939 World's Fair in New York. The centerpiece of this structure was a hundred-foot-diameter turntable display called "The Ford Cycle of Production (see figs. 7.9 and 7.10)." It consisted of eighty-seven separate exhibits, populated by animated figures, all tracing how raw materials were processed into components of a Ford car. The display ascended; fairgoers were meant to look first at the raw materials at the low, outer edge of the turntable and then follow the lines of production as they proceeded upward toward the finished automobiles at the summit.[31]

Ford intended that patrons viewing this extravagant setup learn that putting together one of his fabulous cars required not one but rather two species of wooly domestic mammal. Side by side at the bottom of the turntable were two clusters of figures, one labeled "Sheep herding" and the other designated "Goat shearing." Tracing upward from the shepherd and his charges, the eye moved through the process of carding and dying wool, and then to the finished products of noise insulators and felt washers. Next door, the sightline up from the goat shearing showed spinning and finishing machines, with seat cushions the end product at the top.[32]

The reference to goat shearing deserves explanation. While historically in the United States sheep were the livestock most commonly raised for wool, they were not the only one—certain other mammals also provided fiber. The goats in the Ford display were angora goats, the source of mohair (see fig. 7.11). The angora breed originated in Asia Minor and was first imported into the United States in the mid-nineteenth century. Richard Peters, one of the founders of the city of Atlanta—and coincidentally a grandson of Pennsylvania Judge Richard Peters, who we met in chapter 5—is widely credited as the driving force in establishing angora goats in America. He imported a number of purebred animals, carefully crossing them with common goats to expand the American stock while simultaneously preserving the characteristics of the angoras. Mohair was quickly recognized as ideal

FIGS. 7.9 & 7.10. "The Ford Cycle of Production" at the 1939 World's Fair. This detail shows how sheep contribute to the process. Collections of the Henry Ford Museum of American Innovation, object ID 64.167.232.1704.

FIG. 7.11. An angora goat. Author's photo.

for passenger seats on railroad cars because of its resilience under constant force exerted by passengers' posteriors. As one enthusiast put it in 1876, "The fiber springs back to its original uprightness when the pressure is removed." When automobiles first appeared, mohair found a new use in car seats, and, as the Ford Motor Company made clear at the New York World's Fair, the angora goat was still the source of the raw material for their seat cushions in 1939.[33] Early automobiles were sometimes called "horseless carriages," but Henry Ford's company wanted to make it quite clear that while motorized vehicles could replace horses, those vehicles would still depend on sheep and angora goats. It took not just steel and rubber to build a car but also wool and mohair.[34]

Some who knew him probably thought of Frederick Law Olmsted as a bit of a throwback. Even so, it's easy to picture the Boston park commissioners reading with surprise a detail in the famed landscape architect's 1886 plan for Franklin Park: "To sustain the designed character of the Country Park, the urban elegance generally desired in a small public or private pleasure ground

FIG. 7.12. A lawn mower available when Olmsted designed Franklin Park in Boston. Ad in *Garden and Forest*, February 29, 1888. Library of Congress.

is to be methodically guarded against. Turf, for example, is to be in most parts preferred as kept short by sheep, rather than by lawn mowers."[35] With these words, Olmsted tersely acknowledged the march of technology. He noted two possibilities for keeping the grass cut, but he strongly advocated for ovine jaws as opposed to rotary blades. When he wrote this, mowers had only been an option for a little more than half a century. Coincidentally, the inventor of the mechanical alternative to grazing sheep also built machines used to cut sheep's wool.

English mechanic Edwin Beard Budding was an innovator. One of his notable inspirations was to add a screw adjustment to a wrench, thus creating the adjustable wrench we know today. In the 1820s, Budding built and maintained machines for the British textile industry, including those designed for cross-cutting woolen cloth. He reasoned that if a machine could cut wool in the factory, the same principles could be used to make an apparatus that could cut grass in the field. With a little financial help, he built the first lawn mower, patenting it in Britain in 1830. A new era in landscaping had begun (see fig. 7.12).[36]

Before Budding's machine appeared, scythes were used in some places to cut grass.[37] This was quite labor-intensive. On his 1786 visit to Blenheim Palace in England, Thomas Jefferson reported that there were two hundred groundskeepers for the estate surrounding the palace. "The turf is mowed once in 10 days, in summer," he added.[38] Blenheim was the home of British nobility—the Duke of Marlborough—so resources were available to sustain such a human workforce. But this opulence was uncommon on both sides

Machines in the Pasture

of the Atlantic, which meant that lawns were scarce before the development of mowers. Where expanses of turf did exist, often sheep kept the grass clipped.[39] After Budding introduced the mower, the need for sheep as a landscape maintenance crew diminished.

In New York's Central Park, the nation's first large urban park, the Board of Commissioners report for 1866 boasted that 483 acres of the green space were cut "by the most approved lawn mowers and rollers." There were spots, it was admitted, where mowers were not practical and the grass cutting had to be done with scythes and sickles, but the report insisted that the mower-cut grass looked better than turf cut any other way.[40]

This is not to say that Central Park had no grazing sheep; it did. Just one page after the commissioners praised the efficiency of mowers in their 1866 report, they also noted that, thanks to those mowers, "hay sufficient to supply the sheep and other animals belonging to the Park during the winter ha[d] been made and properly secured."[41] The reason for the qualifying phrase "during the winter" is that in the warm months those sheep sustained themselves grazing in a section of the park, as they had done by that time for several years.

The Central Park commissioners valued the sheep for their wool as well as for their grazing services. Among the assets counted in the 1866 report were $291.33 received from the sale of wool, equal to over $4,700 in 2020. With the flock expanding, their progeny could also be sold; this brought in $425, better than $6,900 in 2020.[42] Clearly the grazing sheep were paying their way—but why, if the powers that ran Central Park were so thrilled with lawn mowers, were the grazing sheep there in the first place?

In answering this question, note that the design of Central Park, like Boston's Franklin Park, was the work of Frederick Law Olmsted, working with his partner Calvert Vaux. One of the specifics of their 1858 plan for the park was an area termed the "Parade Ground." This expanse, their notes on the design specify, would be "a broad open plane of well-kept grass." But, unlike Olmsted's advice to the Boston Park Commission nearly three decades later when he stressed his wish for sheep, there is no indication in the Central Park proposal how that lawn should be well kept. Did he favor it done by humans or by sheep? Whether or not he had an opinion on this, Olmsted did not share it in the earlier of the two park plans.[43]

Nevertheless, sheep would soon come. In its 1864 report, the Board of Commissioners dutifully reported that over the preceding year the park had obtained sixty-three sheep—two African broad-tails, the rest all Southdowns.[44] The following year's report proclaimed: "The flock of Southdown sheep has largely increased and is a very great attraction on the lawn."[45]

But this was not the beginning of the ovine story of Central Park. Clearly there were some sheep on-site even before the commissioners announced the acquisition of a large flock. This is shown, curiously enough, by some harsh remarks in the 1862 annual report about disorderly behaviors by a few park visitors. "While the great mass readily recognize the propriety of reasonable rules," the commissioners asserted, there were those who "chafe[d] against the restraints." Among those rule breakers were those who allowed their dogs to "fly at the sheep and deer that are feeding on the lawn."[46] These sheep may have been strays owned by people living nearby.

As sheep became more of a fixture Central Park, they caused a change in nomenclature. By the twentieth century, the Parade Ground of Olmsted's plan became known as Sheep Meadow, as it continues to be called today.[47] Perhaps it is this transition from a name referencing marching humans to one alluding to grazing ovines that has led to a puzzling modern interpretation. Some sources have stated that the Park Board in the 1860s objected to use of the Parade Ground for military drills and that a primary reason for the introduction of sheep was to discourage this. One historian goes so far to write that the presence of the sheep "effectively prevented troops from conducting drills during the Civil War in Central Park."[48]

The notion that militiamen would be deterred from maneuvers on the Parade Ground by a flock of peaceful sheep is a bit comical. Sheep are not known for preventing humans from either rehearsing for or actually engaging in their periodic violent conflicts. If sheep formed an effective blockade to advancing troops, one might suppose a few wars would have been prevented, or their impact mitigated, by simply massing flocks along disputed borders.

There is a far more reasonable explanation for why sheep were purchased for Central Park, but it has more to do with actual combat in Shiloh and Gettysburg than with practice drilling in New York City. Because of wartime labor shortages, anything that could be done to reduce manpower for park maintenance was welcome—and sheep cutting the grass instead of humans with mowers or scythes helped accomplish this. The impact of the Civil War on operations had been addressed by the Board of Commissioners in its 1862 report. "The effect of [the Civil War] upon the condition of operative industry has been very apparent," they lamented. The report specified that there were fifteen hundred workers employed in 1861; by the next year that number had fallen to nearly twelve hundred.[49] A loss of around 20 percent of the workforce would naturally cause those in charge to seek solutions to the diminished number of employees. Using sheep instead of men to cut the grass was just such a solution.

A review of the timeline gleaned from the Board of Commissioners annual reports supports this conclusion. In 1860, before the war began, the commissioners fretted about stray livestock in the park, including sheep, and noted that these four-legged nuisances were being impounded.[50] In 1862, with the massive conflict raging, instead of complaining about trespassing sheep, the commissioners decried dog owners letting their canine companions charge sheep on the lawn. Two years later, with the bloody war still in force, the commissioners reported that the park had acquired sixty-three sheep to call its own. And then, two years later still, in 1866, with the war finally over and Americans wanting to get back to normal, the commissioners proudly described the use of mowers—not sheep—to keep the grass cut. The sheep were valued precisely when their services would help the Union effort, not so much before or after the war. By simply doing what sheep do—grazing peacefully—the flock supported park maintenance at a time when men were taking up arms and traveling south.

This leads to Frederick Law Olmsted and his request for sheep instead of mowers in his 1886 plan for Boston's Franklin Park. He justified the preference as a means of preserving "the designed character of the Country Park." But his own experiences in park management suggest that something else might have been in his mind.

While Olmsted codesigned Central Park, he also assumed the role of park superintendent, putting him in charge of the workforce that would be entrusted with implementing his blueprints and renderings.[51] He found those workers intractable, and in January 1861 he quit, charging that the men he inherited were "a mob of lazy, wreckless [sic], turbulent & violent loafers." In a brochure written twenty years later, Olmsted even accused the workers of intentionally breaking park equipment with sledgehammers.[52] Having endured such misery generated by human laborers, is it any wonder that Olmsted would prefer a nonhuman option if it was feasible, such as for lawn care?

This is not to suggest that Olmsted was dishonest when he recommended that the Boston park commissioners maintain grass-cutting sheep to provide a more rural appearance. Since he had designed Central Park almost thirty years earlier, Olmsted, like other Americans, had experienced a devastating war and a continuing march of industrialization that was rapidly changing the nation. In this context, it is not surprising that the pastoral-loving Olmsted favored sheep over mowers as a homage to a simpler time.

But when we consider Olmsted's constant struggles against recalcitrant workers in New York, it is easy to imagine that by the time he planned Franklin Park he had concluded that having sheep would mean fewer bothersome humans to deal with. We saw earlier Benjamin Franklin's comparison of sheep and slaves, in which he quipped that an advantage of sheep is that they do not make insurrections.[53] Had Olmsted incorporated that analogy to his time, he might have asserted a similar preference for sheep over two-legged park workers. Sheep will not smash equipment to smithereens with a sledgehammer.

Another prominent American lawn surrounds the White House in Washington, D.C. Like other elegant spans of turf before the introduction of the lawn mower, the grass there was cut by sheep. This practice ended after the Civil War. When Ulysses Grant was president the grounds were groomed by a large mower pulled across the grass by a mule.[54]

There was, however, a reversion to ovine lawn maintenance during Woodrow Wilson's administration. In April 1918, the president went for a drive in the country with Cary T. Grayson, the White House physician. Spotting some sheep, Wilson remarked that both he and the First Lady would like to have a flock at the White House. Dr. Grayson replied that he knew someone who had good stock for sale, and a purchase was quickly arranged. Sixteen sheep—twelve adults and four lambs—were soon brought to the executive residence (see fig. 7.13).[55]

Over the next two years, the press covered the White House sheep mostly as a human interest story. In August 1919, for instance, the *Washington Post* noted with amusement that a conference of a legislative foreign relations committee, held at the White House, was interrupted when sheep gathered just outside the room housing the meeting. The animals proceeded to make so much noise that a policeman had to be instructed to go outside and chase them away.[56]

FIG. 7.13. Sheep on the White House lawn, circa 1918. Library of Congress, digital ID: cph 3a52085.

But the Wilson administration emphasized the serious side of this foray into livestock keeping: again a flock was desired as a wartime expedient. With America engaged in World War I, manpower shortages made it desirable that the White House grass be cut by sheep rather than by humans, repeating the story played out on the Central Park lawn during the Civil War. Furthermore, not only would the jaws of the White House sheep aid in serving America's military effort, their fleece would do the same.

Less than one month after the flock was transferred to their new Washington home, they were sheared, providing over ninety pounds of wool. The wool was auctioned, with proceeds going to the Red Cross war relief fund. In what evidently was an effort to emphasize that the war affected all Americans from coast to coast, there was a separate auction in each state instead of one centralized sale. To enable this, the wool was divided equally into packages of about two pounds and distributed to each state's governor for sale. It seems clear that such a decentralized sale of wool in such small amounts was undertaken with the idea was that Americans would bid far more than the wool's actual worth as a display of charity and patriotism. It worked. Ten months later—after the guns of the war had fallen silent—the Red Cross reported that the auctions had raised about $300 per pound of

Chapter Seven

wool sold, bringing in a total of $30,000, equivalent to over $514,000 in 2020.[57]

With the coming of peace, the sheep remained at the president's residence but not for long. In the summer of 1920, with Warren Harding looking forward to what he hoped would be victory in the fall presidential election, he received his first request for a patronage job. Frank Reece, who somehow had gotten the nickname "Klondike," asked Harding if he could have the position of White House shepherd. Reece was described as "a Civil War veteran, who in most of the years since Appmattox [sic] ha[d] specialized in the difficult art of sheep feeding." (Very likely this should have read "sheep breeding.") The math makes Reece's request rather poignant; even a very young Civil War soldier would have been over seventy by 1920, so this was an elderly man seeking work. And all for naught. The article on Reece's meeting with Senator Harding noted, "It is not known for sure here that the post desired even exists." Very soon it did not. Less than two weeks after the man known as Klondike made his appeal, the Wilson administration announced that the flock, which had grown to forty-eight head, would be sold. Two and a half years after sheep reappeared on the grounds of the presidential residence, they would again lose their job to lawn mowers. Klondike Reece, we can only hope, found some other ovine companions to occupy him in his golden years.[58]

Toward the back of the September 30, 1899, issue of *Harper's Weekly*, there is a half-page advertisement for the Libby, McNeill & Libby Company, urging readers to buy the firm's canned meats. Featured in the ad is a large illustration, but it does not depict of any of the company's foods. Instead, remarkably to our modern sensibilities, the Libby Company thought the best way to tout their product was to show a big, axonometric view of their massive factory in Chicago, complete with two belching smokestacks in the background (see fig. 7.14). "Every can of Libby's Luncheons," the text proudly declares, "is the product of correct cooking, prepared by experienced chefs in hygienic cleanliness in Libby's famous kitchens—largest of the kind in the world." Flanking the factory illustration are two columns listing each of those "Libby's Luncheons." There are more than two dozen of them, including a few that remain midday meal mainstays for some Americans, such as pork and beans and deviled ham. There are also several Libby's Luncheons

FIG. 7.14. Ad for Libby's canned meats (including roast mutton), *Harper's Weekly*, September 30, 1899.

not commonly found in American brown bags today, including ox tails, compressed pig's feet, and—central to the story here—roast mutton.[59]

The bulk of this book centers on sheep in the United States as a source of wool for textiles. Yet there are other uses humankind has had for products from domestic sheep, especially their meat. But while lamb and mutton have long been relished elsewhere, they have not always been favored in America. In 1919 Samuel Cleaver, an Ohio woolgrower who held executive positions with national and local sheep-boosting organizations, declared to the U.S. Tariff Commission: "Dating back, our people were not a mutton-eating people as the English people were.... It was counted cheap."[60] The commission also heard from L. L. Heller, the assistant secretary of the National Woolgrowers' Association. Describing what he perceived as the attitude of his countrymen, Heller said, "There has always been a popular prejudice against lamb. A great many people think that it has an objectionable flavor."[61]

Being in the sheep business, Cleaver and Heller had a vested interest in pushing their product. Accordingly, one could view their comments with suspicion, wondering whether for them it was a case of Americans not eating sheep or simply that they were not eating *enough* sheep to satisfy their business goals. The statistics Heller provided in his testimony mostly support the latter conclusion. America's annual per capita consumption of lamb and mutton, he reported, stood at about seven pounds. That compared to seventy-one pounds of beef and sixty-seven pounds of pork. While those numbers show that cattle and hogs were far more important than sheep as food animals in the United States, this still meant that in 1920 over 742 million pounds of lamb and mutton were being served at the nation's dining tables.[62] Indeed, time and again the Tariff Commission heard in its 1919 interviews with sheepmen that meat was now an equal or greater source of revenue to them than wool.[63]

But Cleaver and Heller weren't only bothered that Americans didn't eat mutton and lamb as frequently as beef and pork. Heller also protested that his countrymen did not consume sheep meat as much as their English brethren who, he declared with envy, annually ate about twenty-six pounds of mutton and lamb per capita.[64] Others have also commented on this difference between Americans and British and mused that it is surprising in light of America's British heritage. "Knowing the propensity of the Englishman for sheep flesh, one wonders why (or if) English colonists lost their tastes for lamb and mutton so quickly," remarks a thorough survey of food supply in the antebellum South.[65] A mid-nineteenth-century sheep husbandry manual observed, "Mutton and lamb are a favorite, if not *the* favorite food of the English of all classes. . . . On the other hand, it is evidently *not* a favorite meat in the United States (emphasis in original)."[66]

But it wasn't a loss of *taste* for sheep or any intrinsic difference between Americans and Britons that led to what we might call "the mutton gap." The history of American sheep provides a number of reasons why mutton was less common for dinner on this side of the Atlantic. First, because of their susceptibility to predation, sheep were not well established in colonial times or in the early decades of United States, and thus there were simply fewer of their kind to butcher than was the case with cattle and swine.

Closely related to this, the scarcity of sheep, coupled with the desire to reduce or eliminate the need for wool imports, meant that colonists and early Americans valued the sheep far more as a source of clothing than of food.

Recall that the First Continental Congress pledged to "improve the breed of sheep" and that to do so they would have to "kill them as seldom as may be."[67] It was thus a matter of honor and patriotism for sheep owners to keep all their stock alive for the shearers to fleece.

Finally, the favored "improved" breeds of sheep in the early nineteenth century received mixed reviews as to the quality and flavor of their lamb and mutton. Some insisted that Merino flesh was tasty, but many disagreed, leading to a strong perception that the breed had superb wool but inferior meat.[68] Mutton from Saxony sheep was widely despised.[69] Given these handicaps, it isn't surprising that beef and pork came to be more prevalent on the American menu than mutton.

Consumption of sheep meat was not entirely alien to early America. That it was sometimes relished is apparent from a look at the pages of *The Virginia House-wife*, by Mary Randolph. With a first edition appearing in 1824, *The Virginia House-wife* was possibly the first cookbook published in the United States, and it has been called the most influential cookbook of nineteenth-century America. The book includes four ways for preparing lamb and thirteen for mutton. In a recipe for lamb that still sounds delicious, Randolph advises the reader, "The fore quarter should always be roasted and served with mint sauce in a boat, chop the mint small and mix it with vinegar enough to make it liquid, sweeten it with sugar. The hind quarter may be boiled or roasted and requires mint sauce; it may also be dressed in various ways." Mutton, she insisted elsewhere, "is in the highest perfection from August until Christmas, when it begins to decline in goodness." Randolph also cautioned that while a sheep's liver "is very good when broiled," the neck of a sheep "is fit only for soup."[70]

But there was an unavoidable limit to the consumption of lamb and mutton—or beef and pork, for that matter—plaguing people in the preindustrial age. Meat from slaughtered animals soon spoiled from the elements, putting temporal and spatial limits to safe human consumption.[71] By the late nineteenth century, however, advances in technology allowed the mass slaughter of livestock for meat that would not be eaten presently or nearby. The Libby's factory so prominently pictured in the company's 1899 *Harper's Weekly* ad was in Chicago. That location was not happenstance.

The poet Carl Sandburg famously called Chicago "Hog Butcher for the World."[72] He might just as accurately have dubbed the city "Cattle Butcher for the World" or, indeed, "Sheep Butcher for the World." None of these superlatives would have been an exaggeration. In 1896, eighteen years before Sandburg published his poem "Chicago," one glowing account combined the city's involvement with swine, beef, and mutton to declare that Chicago had become "the greatest live stock market in the world."[73]

With a large number of livestock slaughtered by different companies, it quickly became apparent that it was preferable to have a central location for this, rather than for each meatpacking firm to maintain its own slaughterhouse. It was also seen as beneficial to all companies for there to be a common location for receiving livestock to butcher. And so the Union Stockyards on Chicago's South Side were born in 1865 as a replacement for a smaller facility. By 1900 these stockyards covered 475 acres.[74] The complex grew as a result of Chicago's position as a railroad hub linking all corners of the nation. Trains from the western states would pull into the station adjacent to the stockyards where cattle, hogs, and sheep were offloaded. In 1895 alone, the stockyards received shipments totaling over fourteen million livestock. Of these, 3,406,739 were sheep, meaning that ovines represented nearly a quarter of the animals brought to the Chicago operation.[75]

Once off the train, livestock had to be quartered before being transported and killed. On any given weekday in the 1890s, workers at the stockyards slaughtered, on average, fifteen thousand hogs, five thousand cattle, and five thousand sheep.[76] At the Swift and Company plant, six hundred sheep per hour were slaughtered with such efficiency that it took only twenty-six minutes for a live animal to be killed, dressed, and turned into cuts of mutton whisked into the cooler.[77] That there were by that time coolers was central to successful operation of the stockyards. The establishment of a single huge site in the Midwest to handle so much meatpacking was only feasible due to the development of refrigeration. Prior to this, livestock were slaughtered near where their meat would be sold, spoilage being a significant problem. But the 1870s saw the invention of refrigerated railroad cars and shipping holds that allowed livestock to be slaughtered in Chicago and then the meat safely transported to consumers all around the United States and even overseas.[78] Bolstered by the new technology, exportation of fresh mutton began in 1877. Over 99 percent of this went to the United Kingdom and Ireland. By 1880, a scant three years after exports began, over 2.3 million pounds of American mutton arrived on British and Irish shores.[79]

But a sheep carcass is more than just edible flesh, so it was good business for the meatpacking companies to find productive uses for all its parts. The

FIG. 7.15. Ad for wool soap in *Harper's Weekly*, September 30, 1899.

same issue of *Harper's Weekly* carrying the ad for Libby's canned mutton also has an ad for a by-product of slaughtered sheep. This ad features an illustration of two happy little girls, about three or four years old, one of whom blows a bubble that the other prepares to try to catch in her cupped hands (see fig. 7.15). "Blowing Wool Soap bubbles is great fun," the copy accompanying the picture notes, "but it is greater fun to know that by using Wool Soap for all home purposes, especially for the toilet and bath, it means a saving in household expenses." A line at the bottom of the ad states that Wool Soap is a product of Swift and Company—the corporation that slaughtered six hundred sheep an hour. Swift primarily focused on food, and the Libby Company that produced the canned roast mutton was one of its subsidiaries. As the soap ad attests, however, Swift branched out from the kitchen to the bathroom.[80] An October 1899 issue of *Harper's* carried another ad for the soap, this one highlighting its substance. "Wool Soap," the ad asserted, "is made from the purest fats—the best grade of pure mutton tallow."[81] Thanks to Swift's operations at the Union Stockyards, tallow was available for conversion into a gentle soap, fine for the skin, safe for clothing, and superb for bubble blowing by little girls. Sheep tallow was also employed in candle making.[82]

Another ovine product is catgut, which comes from sheep intestines. Surgeons valued catgut for the sutures vital to their practice.[83] For tennis players, this by-product of slaughter was essential in its use as strings for their rackets.[84] It was musicians, however, who truly had strong views about catgut.

Violinist Nathan J. Landsberger dined with a group of fellow musicians at a San Francisco café one Saturday night in January 1914. According to an account in the *San Francisco Examiner*, Landsberger was near tears as he wailed that Italian peasants were carelessly allowing their lambs to eat tin cans.[85] It wasn't concern for animal welfare that upset him. Rather, he was annoyed by what this consumption of metal was doing to music. As he told it, the best violin strings had always come from Italy, and he asserted that the excellence was due to the loving care of Italian shepherds. Nobly and carefully, he insisted, they looked after their lambs, which "were fostered and raised for the sole purpose of having their little insides later made into violin strings." These strings, Landsberger said, "enthralled the world."[86]

But those days were long gone. Italian peasants had become modernized, and an unfortunate side effect of this was that they weren't as attentive to their lambs, which were Landesberger maintained, running about swallowing pebbles in addition to ostensibly eating cans. Thus "their little insides were full of punctures," and this translated into defective violin strings that would snap "when a musician [was] deep in a Chopin masterpiece." If the Italian shepherds did not go back to their old vigilant ways, Landsberger lamented, the violin was doomed.[87]

One gets the impression from the article that Landsberger believed all this. Less clear is whether the anonymous author wrote tongue in cheek, especially when suggesting in conclusion that American tourists in Italy should gather the shepherds and "tell them that pebbles are not good for little lambs that are going to be made into violin strings."[88] While Landsberger's opinion about violin strings may have been extreme, other musicians were also quite particular about catgut and strings for their instruments.

The origin of the word "catgut" is unclear. An archaic definition of the word "kit" dating from the sixteenth century is "a small violin," and sheep gut strings on a kit could have morphed into "catgut."[89] Another etymology notes that an early manufacturer of strings for instruments was located in Catagniny, Germany, and that "Catagniny gut" got shortened to "catgut."[90]

Violinists and violin makers have been less concerned with this etymology than with the quality of catgut, as those who dined with Nathan Landsberger could attest. In 1925 the violinist and composer Alberto Bachmann published an encyclopedia about his favorite instrument. He stressed that sheep useful for violins had to be from just the right sort of habitat, and he insisted that strings of the highest caliber came from animals "raised on dry pastures and in hilly countries."[91] Moreover, he wrote, while some string makers bought their sheep intestines from butchers, most would rather not do so, because how could they be certain the guts were fresh? Instead, according to Bachmann, those string makers preferred "to extract them from the bodies of the sheep while still warm, or to have this done by workmen experienced in so doing." Otherwise, "they would risk receiving them in a state of alteration which would make them unfit for the purpose."[92] Beautiful music, Bachmann maintained, was dependent on freshly killed sheep.

But the matter went deeper. Bachmann approvingly quoted a violin maker who asserted that a different sort of sheep was needed to provide the E string—the thinnest one. An E string should not be too white, the maker declared, because that would mean it had been "made of a gut of too young a lamb." On the other hand, he advocated that the other strings should be *very* white, which by his formulation meant coming from a young lamb that had not yet aged to a point where it could provide a proper E string.[93] Knowing the age of the slaughtered animal, in other words, was key to determining whether its guts should be turned into E strings or A strings. The same sheep carcass could not provide both.

While Nathan Landsberger praised strings made in Italy—at least those produced before the alleged epidemic of can-consuming lambs, with five thousand sheep processed daily at the Union Stockyards it would not be long before someone decided that Chicago would be a natural home for manufacture of strings for instruments. In 1899 two English immigrant brothers set up a small shop near the stockyards. Interviewed by a local newspaper, one of the brothers said he had wanted to make strings twenty years earlier but had been thwarted by the lack of enough fresh carcasses to make business profitable. That changed when Chicago became a center for sheep slaughter. In 1899 he was glad to report, "The material is readily obtainable and fresh."[94]

FIG. 7.16. Armour employees cleaning sheep intestines to be made into strings for musical instruments, circa 1917, *The Violinist*, April 1917.

The nascent factory depended on the local meatpacking giants Swift and Armour for a supply of intestines, but this business model had a problem. One of the employees at Armour & Company, Harley O. Gable, pointed out to company executives that instead of selling their sheep entrails to firms that manufactured them into strings, it would be profitable to make the strings in-house. Management agreed, and in 1912 Armour began a new business catering to musicians, with Gable in charge (see figs. 7.16 and 7.17).[95] As most strings in the past had been made in Europe, the company evidently reasoned that their new product should be marketed with Old World–sounding names, thus Armour's ads in trade publications touted their "Helmet," "Il Trovatore," and "La Melodia" lines of violin strings. All their lines, they assured musicians, were better than German strings and equal to French and Italian ones. To look at it another way, Armour promised that the intestines of sheep that scampered across the foothills of Montana produced violin strings as sublime as those originating in the guts of sheep frolicking over the hills of central Europe.

Machines in the Pasture

FIG. 7.17. Ad for Armour violin strings in *The Violinist*, June 1917.

Skillful musicians are usually very particular about the condition of their instruments, and David Mannes was no exception. Born in New York City in 1866 to German immigrants, Mannes had a long and distinguished career as a violinist, conductor, and music teacher.[96] In an interview published in 1919, he was asked whether he preferred strings of gut or of wire. He left no doubt that he believed the sweet tone of a fine violin depended on sheep. Mannes reasoned that Stradivarius and the other old master violin builders had specifically crafted their instruments to be played with gut strings, ovine intestines being the material available in their time. If they had conceived of using metal wire strings, Mannes insisted, they would have created an instrument totally different from the violin as he knew it. Wire strings, he protested, changed the violin's tone, much to its detriment. He acknowledged that wire strings were more durable and "make things easier technically." But he would forego those advantages in order to produce the superb sound he felt only possible with gut strings. Mannes concluded, "When materially things are made easier, spiritually there is a loss."[97]

Catgut violin strings sound divine in the hands of an accomplished musician like David Mannes, but they have drawbacks. Especially problematic in the early history of the instrument was the G string, the thickest of the four. The G is tuned to a low pitch, but gut strings so tuned are not thick enough in relation to their tension to produce a harmonious sound. In 1765, this obstacle was first addressed when the French composer Jean de Sainte-Colombe wound copper around a gut G string, increasing its mass. Later other metals, such as silver, copper, bronze, and aluminum were used to wrap not only the G, but also the D and A strings.[98]

As the thinnest string, the catgut E had a different problem: it broke easily. (An E was probably the string Nathan Landsberger cursed for snapping in the middle of a Chopin masterpiece.) Further, because E strings were so fine, wrapping them with wire as was done with the G was out of the question. Given this situation, a purely metal string would be desirable, but manufacturing them would require great precision. Not until the late nineteenth century was production of steel E strings technologically possible.[99] By the time of David Mannes's 1919 interview, violins with all four strings wholly or partially steel instead of catgut were common, and thus a question of preference was put to him that couldn't have been asked of earlier generations of violinists. But even among musicians who might have preferred strings originating as sheep intestines, often a switch to metal took place for economic reasons. Steel strings were simply less expensive.[100]

While David Mannes was firm in his preference for catgut violin strings, he was even more resolute in his belief that the musician's art hugely benefited humankind as a wellspring for peace. As he put it: "The only times when I have witnessed a state approaching the brotherhood of man have been moments of music, when hundreds of hearts beat to the same rhythm and lifted to the same phrase, and when all hate, all envy, all greed were washed away by the nobility of sound. Words are so often the agents of destruction; music—good music—can only build."[101]

When Sheriff Patton set out in Casper, Wyoming, that June day in 1918 to make certain that local shearers were shearing and not striking, probably the farthest thing from his mind was that a sheep's intestines could be fabricated into superb violin strings, agents of sublime music. From that same animal, one part could be employed to clothe armies of destruction, while another part, as David Mannes would wish, enabled the gentle sounds of a Tchaikovsky violin concerto. In the long history of the relationship between humans and sheep, sheep have bountifully provided for our best instincts and our worst, as a source of raw materials for both war and peace.

Machines in the Pasture

CHAPTER 8

Nothing Is Certain but Death and Taxes on Wool

Foreign Sheep and Internal Revenue

> In the whole history of the tariff, probably no schedule has excited more attention and controversy than the wool and woolens schedule.
> —CHESTER WHITNEY WRIGHT, *Wool-Growing and the Tariff*

In December 1887, when President Grover Cleveland delivered the third annual message of his first term, he did not speak from a raised podium in front of a joint session of Congress. Instead, he followed the practice of the time, writing a report that was transmitted to the legislative branch.[1] But even with Cleveland's thoughts expressed on paper, his somber tone was apparent. The nation's finances, he said, were in sorry shape: "The amount of money annually exacted, through the operation of present laws . . . largely exceeds the sum necessary to meet the expenses of the Government."[2] Cleveland was complaining that the U.S. Treasury had too much money— its "surplus revenues" had that very month topped $55 million and was forecast to keep rising![3]

To Cleveland this was unacceptable. In America, he argued, every citizen was entitled to "the full enjoyment of all the fruits of his industry and enterprise, with only such deduction as may be his share toward the careful and economical maintenance of the Government which protects him." For the federal government's taxes to take any more than this was "indefensible extortion and a culpable betrayal of American fairness and justice." Cleveland insisted that the money sitting idle in the Treasury belonged in the hands of Americans to purchase goods or to start or expand businesses.[4]

Part of the problem, the president felt, was that the wool tariff was too high. A statesman advocating for or against a policy often will elaborate on how the practice affects ordinary people. Cleveland's message would be no exception, and the ordinary Americans he pleaded special empathy for were sheep farmers. Cleveland noted, "The farmer ... is told that a high duty on imported wool is necessary for the benefit of those who have sheep to shear, in order that the price of their wool may be increased." Initially, the farmer might be quite pleased, Cleveland conceded, because he would have more money in his pocket from selling his wool at a higher price.[5]

But, the president argued, that attitude was short-sighted, as that sheep farmer would learn when he himself became a consumer. A high proportion of the nation's sheep, the president commented, were kept in small flocks of twenty-five to fifty animals. "[When the small farmer] finds it necessary to purchase woolen goods and material to clothe himself and family for the winter," Cleveland declared, "he discovers that he is obliged not only to return in the way of increased prices his tariff profit on the wool he sold ... but [also] that he must add a considerable sum thereto to meet a further increase in cost caused by a tariff duty on the manufacture." This meant that the sheep farmer's pocketbook was worse off than it would have been without the high tariff allegedly protecting him, and so, according to Cleveland, the current tariff was "a tax which with relentless grasp [was] fastened upon the clothing of every man, woman, and child in the land."[6]

Cleveland carefully explained why a high tax on imported wool garments hurt even those who purchased only domestic clothing. Yes, he admitted, comparatively few people purchased the imports. But because the tariff kept the price of foreign wool goods high, the domestic manufacturers were selling their wares at an inflated cost, one near or equal to what consumers would pay if they had bought woolens from overseas that were subject to the duty.[7] In other words, he was arguing that there was an artificial economic situation played out to the detriment of ordinary Americans. In the absence of the tariff the price of the imported fabrics would be lower, and American manufacturers would correspondingly be forced to also lower their prices so not to lose business to foreign interests. The consumer, by Cleveland's thinking, would thus benefit.[8]

According to the Constitution, the president uses the State of the Union message to recommend to Congress "such Measures as he shall judge necessary and expedient."[9] A president proposing new legislation is obviously disadvantaged if either House is controlled by the opposition party. That was the situation facing Cleveland. While the majority in the House of Representatives were, like him, Democrats, the Senate had a Republican majority.[10] And so Cleveland did not get his bill lowering wool tariffs. While Cleveland vehemently argued that high tariffs hurt sheep farmers, his Republican rivals just as forcefully proclaimed that high tariffs *helped* American woolgrowers.

The Republican Party's origins are commonly associated with the "free labor" ideology and the antislavery movement. But throughout the nineteenth century, Republicans promoted another cause: making sure that imported wool was subject to a hefty duty. This dichotomy between antislavery and wool tariffs was analyzed in a paper by political scientist Christopher H. Achen.[11] He noted that in 1850 Vermont became the most Republican state in the nation, and, since it was also the first state to prohibit slavery, some have assumed this was what drew voters so heavily to the Republican side. But Vermont at that time was a leading wool-growing state, and Republicans were for a wool tariff, so others have argued that Vermont became largely Republican due to sheep farmers voting for their own economic self-interest. Achen wished to determine which theory matched best with the facts.[12]

From analyzing a variety of data, especially voting patterns and agricultural censuses, then correlating the similarities between them—that is, how did residents in the counties having the most sheep vote compared to those of counties with the fewest sheep—Achen concluded that neither antislavery sentiment nor pro–wool tariff sentiment were by themselves sufficient to explain Vermont's mid-nineteenth-century politics. Rather, he said, "it was their dynamic interaction . . . that led to the Republican realignment."[13] It seems logical that Achen's interpretation would hold for other free states with large sheep flocks.

In any case, Republican support for wool tariffs remained high even after the party's antislavery position became moot with the Union victory in the Civil War. So President Cleveland's 1887 plea to lower the tariff had to be met head-on. Cleveland's tariff position had been known three years earlier when he had successfully run for president, leading the Republicans in their 1884 platform to emphasize their concern for sheep farmers. "We recognize the importance of sheep-husbandry in the United States, the serious depression which it is now experiencing and the danger threatening its future prosperity," the platform declared. Because of the value of American sheep, the Republicans promised to support "a readjustment of duty upon foreign wool in order that such industry [should] have full and adequate protection."[14] "Readjustment" meant they wanted a higher tariff.

The year after his 1887 annual presidential message, Cleveland faced re-election, opposed by Benjamin Harrison. Accepting the Republican presidential nomination in 1888, Harrison asserted that substantial duties on foreign wool gave American sheep farmers "protection for their fleeces and flocks, which ha[d] saved them from a further and disastrous decline." Horrified that the Democrats favored duty-free wool, he added, "The people of the West will know how destructive to their prosperity [this] would be."[15]

When Cleveland and Harrison faced off in 1888, historian Joanne Reitano argues, they were competing for the vote in an election where the tariff "was at the center of the American mind" in a way it had never been before nor has it been since.[16] Thomas Jefferson intimated to Lafayette in 1813 that it was worth armed conflict to spur an increase in American efforts to improve and expand flocks of Merinos.[17] If the War of 1812 can be thought of in part as a war about sheep, the presidential election in 1888 might as plausibly be regarded as a referendum about sheep.

Earlier we saw how sheep were part of the congressional debates on the "Tariff of Abominations." That 1828 bill was hardly unique; during the nineteenth century, Congress passed seventeen acts adjusting the duties on wool.[18] This is not the place to examine them all, but it is impossible to discuss the importance of sheep in American history without touching on wool tariffs, since these were a significant portion of the revenue collected by the U.S. government in the 1800s. For example, in the 1887 fiscal year, shortly before Grover Cleveland's message to Congress, the Treasury took in over $371 million in total revenue. This included nearly $5.9 million collected in duties on the import of raw wool and a further $29.7 million in duties on imported clothing and other goods manufactured from wool, making the total of wool-based tariffs $35,629,534—nearly 9.6 cents of every dollar.[19] As a comparison, when the federal government took in $3.33 trillion in fiscal year 2018, the tax on gasoline contributed $36 billion. Accordingly, gas taxes were only about 1.1 cents of every dollar of revenue.[20] Thus, the revenue chain beginning in 1887 with a sheep being sheared in Australia, Argentina, or elsewhere provided the federal coffers with a much greater percentage of the operating revenue for the U.S. government than is provided by modern American motorists filling their gas tanks.

Why the appeal for raising revenue by taxing wool imports? Arguably this was less a matter of economics than of politics. Frank Taussig, looking at the subject in 1910 from the perspective of an economist who had lived through decades of debate on wool duties, wrote that wool was "almost the only article through which protection could be promised and given to agricultural voters." Tariffs were also laid on wheat, corn, and meats, but, unlike wool, imports of these were dwarfed by exports. Thus, Taussig asserted, wool tariffs meant that "the farmers got some share of the benefits of the protective system" to a degree that they would not get from anything else they raised. Taussig further elaborated that some of the states with the most sheep farmers were closely divided in politics, especially, he noted, Ohio. For this reason "it required some courage among the Democrats to present themselves squarely on the platform of [duty-free] wool."[21]

That is what hurt Grover Cleveland's bid for reelection in 1888. He beat Benjamin Harrison in the popular vote, 48.6 percent to 47.8 percent, but Harrison won in the Electoral College, gaining 233 votes to Cleveland's 168.[22] Cleveland lost seven of the eight states that according to the 1890 census had over a million sheep, and in three of them—New York, Indiana, and Ohio—his margin of defeat was so small that it is probable those state's shepherds were crucial to Harrison's triumph (see table 8.1).[23] If Cleveland had received just 1,175 more votes in Indiana, he would have taken its fifteen electoral votes, and if he had gotten 7,187 more votes in New York, he would have claimed that state's thirty-six electoral votes, giving him a victory over Harrison in the Electoral College, 219 to 182. In an election where over 11.3 million ballots were cast, a total of 8,362 votes in two sheep-rich states decided the election. This is less than one-tenth of 1 percent of the total votes registered. Interestingly, Indiana and New York were the home states of the two candidates.

This is not an economics textbook and thus not the place to consider the pros and cons of the two positions: the Republican assertion that high wool tariffs helped American sheep farmers or Cleveland's belief that those duties were exorbitant and ultimately hurt sheepmen. Near the conclusion of his long 1910 treatise on wool tariffs, Chester Wright argued, "The tendency to overestimate the power of the tariff, either for good or for evil, is almost universal."[24] So, by Wright's estimation, the disagreement between Cleveland and the Republicans was one of little substance.

But there is another relevant point. One need not be an economist to understand human nature, and in this regard it seems clear that Grover

TABLE 8.1. Results of 1888 presidential election in three states with over one million sheep

State	Votes for Harrison	Votes for Cleveland	Margin of victory	Electoral votes	Number of sheep (from 1890 Census)
Indiana	263,361	261,013	0.437%	15	1,081,133
New York	650,338	635,965	1.089%	36	1,528,979
Ohio	416,054	396,455	2.332%	23	4,000,720

Cleveland had a much tougher policy to sell than his Republican rivals did. The Republicans said to America's sheep farmers "We are giving you a high tariff so that you can sell your wool for a greater price." Cleveland, on the other hand, was cautioning the sheepmen to ignore the immediate pleasure of getting more money for their wool. Instead, he stressed, they should remember that eventually the extra income wouldn't matter, because they would themselves pay more as consumers. The Republicans, in effect, were happily offering candy as an immediate gratification, while Cleveland sternly warned that eating too much candy leads to a painful stomachache later. This put Cleveland at a disadvantage.

In spite of the 1888 setback, Cleveland wasn't finished. In 1892 he ran once more, again facing Benjamin Harrison. This time the Republicans were even more blistering in their criticism of Cleveland than they had been four years earlier. In their campaign literature, the Republicans snarled that Cleveland's 1887 speech was "the first annual message by an American President ever devoted exclusively to an attempt to destroy an important industry of his own country."[25] That industry, they emphasized, began with the sheep-raising farmer, who would be put out of business if the Democrats had their way. The argument then drifted into xenophobia:

> [Duty-]free wool means for more than a million American wool growers direct and unchecked competition of the most degrading sort, not with free men of any race who own their lands and support their own government, but with convicts or coolies in Australasia, with the savages of South America and the blacks of South Africa. Where land costs nothing, where the climate reduces the expense of sheep raising to the minimum, and where the laborer has neither home nor citizenship, and lives on nothing, wool can be grown at a cost with which American farmers cannot compete, and should never be obliged to compete.[26]

Notwithstanding the Republicans' appeals, Cleveland won in 1892, returning to the White House after a four year absence.[27]

That rival candidates competed in two consecutive presidential elections, with split results, underscores the difficulties faced by nineteenth-century politicians advocating either an increase or a decrease in wool duties. These tariffs often operated in a manner not predicted by their advocates or by their detractors. This was especially true when national or international events upended the tariff's revenue goals. For example, the efficacy of a wool tariff enacted late in the Civil War was compromised by the coming of peace. Surplus military clothing was then thrown on the market at rock-bottom prices, greatly reducing the immediate demand for wool of any kind, domestic or foreign.[28]

Inconsistency was also a problem. Duties were altered by Congress so often in the nineteenth century that there was never a set policy for long. The ensuing uncertainty, some argued, was detrimental to America's economy. During yet another debate on wool tariffs in 1894, exasperated California representative William Bowers argued that the mere guarantee of no change in the tariff laws for the next twenty years would lead to economic prosperity. He added facetiously that hanging anyone who proposed a rate change would serve to accomplish this. Indeed, the seesaw nature of the tariff legislation in that era is almost comical. An 1867 act carefully spelled out a complex system of duties, depending on the type of wool. In 1872 all these duties were reduced by 10 percent, but three years later the 1867 rates were restored. The intricate tariffs established in 1890 were reduced in 1894, with raw wool being made duty-free, but the 1890 schedule was largely reinstated in 1897.[29] This back and forth was largely a function of changes in congressional majorities due to election results.

Congressman Bowers was a Republican. But feelings that the constant vacillations in tariff laws damaged markets were not confined to his party. Three weeks after Bowers's speech, Benton McMillan, a Democrat from Tennessee, took the floor and declared, "If with each deficiency or surplus Congress attempts to raise and lower duties . . . there can be no stability in values, manufactures, or business." These perpetual rate changes and corresponding fluctuations in value, he added, "produce more trouble than can be produced any other way."[30] McMillan, like Grover Cleveland with his example of the sheep farmer paying more for clothing, also argued that tariffs were a consumption tax, especially burdensome to those with few financial resources. Clearly, McMillan and others thought, the money collected by the government through wool tariffs should be procured in a different manner. But how?

Missouri Democrat James Beauchamp Clark—"Champ" was his nickname—had a long career in Congress, including being elected Speaker of the House from 1911 to 1919.[31] As a freshman in the House in 1894, he caused a stir when he proudly uttered a prediction of what would be accomplished that session. "We are going to impose such an income tax as will make the eyes of the multi-millionaires bulge out," he promised.[32] According to the *Congressional Record*, this bold statement elicited laughter and applause, although one might guess that there were also a few sneers. When the commotion died down, Clark's next words were about what he called "this sheep question." He declared, "There is one-half of a sheep to each man, woman, and child in the United States," adding that the average sheep provided four pounds of wool per shearing. Much like Silas Wright promoting the Tariff of Abominations sixty-six years earlier, here was a member of Congress advocating a controversial tax and, in his speech doing so, lecturing his colleagues on the state of the nation's flocks.

Clark promised the "eye-bulging" income tax in response to a concern raised by a member of his own party, Thomas A. E. Weadock of Michigan. Weadock was skeptical about the wisdom of lowering wool tariffs, saying he was "very much amused" at all the talk of duty-free wool when those supporting it did so "without saying a single word as to how the $400 [million] necessary to pay the expenses of Government are to be provided."[33] Clark's retort that the lost revenue should be raised by taxing the earnings of the wealthy, followed by his discourse on the weight of an average fleece, shows that from the start, the idea of an income tax was closely associated with the status of America's own flocks and its dependency on foreign ones.

Nor was Clark alone. After discussing the importance of sheep to his state's economy, Wyoming Democrat Henry Coffeen asserted that there ought to be "a proper revenue tariff on wool," but he also acknowledged the point made by President Cleveland several years earlier, that a lower tariff would make woolen clothing less expensive—by his calculations, the reduction then being proposed would save the average Wyoming family fifty dollars a year. That, Coffeen happily said, would be a great benefit for "the poorest and humblest of our citizens." Echoing Clark, he said that the deficiency in revenue should be made up through an income tax. Speaking of his state, with its seventy-five thousand people (and over seven hundred thousand sheep), he declared, "We do not believe that it will cripple or ruin the rich men of our country whose incomes are yearly over $4,000 if they shall be called upon to pay 2 per cent tax on their incomes above that amount."[34]

Like Coffeen, Michigan Republican David Aitkin connected the wool tariff with the proposed income tax, but he came to the opposite conclusion. How, he demanded, does the combination of duty-free wool and a corporation paying 2 percent of its profits to the government help the sheep farmer? Aitkin insisted that the sheepman would still take a loss on the sale of his wool, and banks, subject to the new tax, would simply pass it on to those seeking loans. And so, he concluded, farmers borrowing money would be "compelled to help make up this tax in addition to their loss on wool and sheep."[35]

The United States had first levied a tax on incomes in 1862. While it was adopted as an expedient to provide revenue to pay for the Civil War, the tax remained in force several years after Appomattox, finally expiring in 1872. The idea of funding the government through a progressive assessment on the earnings of wealthier Americans, however, continued to have a great deal of appeal. There were protests when the 1862 tax expired, and during the 1870s several bills were introduced in Congress to reinstate it.[36] These failed to garner sufficient support, but Clark's 1894 promise proved true, aided by a Congress that for the first time in a decade and a half featured a Democratic majority in both houses. A revenue bill was passed, including a tax on incomes.[37] The sixty-one-page law, called the Wilson-Gorman Tariff Act, featured long schedules setting new, generally reduced rates for many items, including wool. But it was a short paragraph, section 27 of the act, that enacted Congressman Clark's boast. It provided that all "gains, profits and income" were subject to "a tax of two per centum on the amount so derived over and above four thousand dollars"—exactly the amount endorsed by Henry Coffeen.[38] Such a high exemption (for its time) and such a low rate meant that this truly was a modest fee; an individual making exactly

$1,004,000, an impressive sum at the time, would pay only $20,000 under the new tax.[39] Compared to the rates of our time it scarcely seems eye-bulging, to use Clark's phrase.

But it would not last. The following year, a challenge to the income tax portion of the statute was heard by the Supreme Court. By a five to four vote, in *Pollock v. Farmers Loan and Trust Company* the new tax was struck down, the bare majority arguing that Congress had exceeded its constitutional powers.[40] The decision was roundly criticized by the four dissenting justices—as well as by most modern legal commentators—who accused the *Pollock* majority of both narrowly reading the taxing authority of Congress and of disregarding precedents that would have held the tax valid.[41]

Nevertheless, a ruling by the Supreme Court stands no matter how slim the majority, and so, where revenue collection was concerned, it was back to the drawing board—and, unsurprisingly, back to a higher wool tariff. Congress raised it again in 1897.[42] Support for an income tax remained strong and bipartisan, however. In his inaugural address of 1909, President William Howard Taft—a Republican—noted that the financial panic two years earlier had reduced government revenue from customs dramatically. Thus, he said, "The expenditures for the current fiscal year will exceed the receipts by $100,000,000."[43] Taft argued that a new source of revenue was needed, and just a little more than three months later he sent a special message to Congress, specifically recommending that they approve a constitutional amendment expressly overturning *Pollock* and then send it on to the states for ratification.[44] This happened rather speedily. The Sixteenth Amendment, giving Congress "power to lay and collect taxes on incomes, from whatever source derived[,]" was officially ratified in February 1913, less than four years after Taft's urging.[45] A constitutional amendment must be approved by at least two-thirds of both houses of Congress plus a minimum of three-fourths of the states. That the Sixteenth Amendment achieved this in a short time again demonstrates the popularity a tax on incomes had at the time.[46] Fittingly, the Speaker of the House at the time of the amendment's ratification was Champ Clark, who nearly twenty years earlier had asserted that an income tax was necessary, right before he compared America's human and sheep populations.

Congress applied its new authority almost immediately. A bill was passed and signed by President Woodrow Wilson exactly eight months after the

Sixteenth Amendment's ratification.[47] The title of the legislation—"An Act to Reduce Tariff Duties and to Provide Revenue for the Government, and for Other Purposes"—makes it clear that once again this was largely concerned with taxes on imported goods. There was a long schedule regarding duties on woolens—for example, yarn made of wool would be subject to an 18 percent ad valorem rate, while most imported woolen clothing would be assessed a 35 percent ad valorem tariff.[48] But there would be no such charge on the unmanufactured fleeces directly off a sheep's back. The law specified that "wool of the sheep" was duty-free.[49] Only after the exhaustive, fifty-two-page customs schedule did the statute spell out the specifics of the income tax. As with the aborted 1894 measure, this was to be a modest tax on only the wealthiest Americans. There was a $3,000 exemption for individuals, nearly $80,000 in 2020, and this, coupled with other deductions, meant that only about 1 percent of American adults paid the tax.[50]

The 1913 bill is often called the Underwood-Simmons Tariff Act after its sponsors. Oscar Underwood served Alabama as a member of the House of Representatives for nearly two decades and also as a senator for two terms.[51] Underwood was mentioned as a candidate for president in 1924, but he condemned the Ku Klux Klan in a speech, thus destroying much of the support he might otherwise have secured in his native South and ending any hope of his receiving the Democratic nomination. For this action, he was praised in John F. Kennedy's *Profiles in Courage*.[52] Underwood's detractors, however, did not think him courageous in May 1913 when the bill that bears his name was debated. Instead, he was accused of duplicity. Why, he was challenged, did you just last year support a 20 percent tariff on raw wool when you are now advocating that wool be duty-free?

Underwood explained simply that he had not wanted a wool tariff in 1912, it was just that the government then needed the money that tax would bring. But now, he argued, the ratification of the Sixteenth Amendment rendered wool duties unnecessary. "We did not have the right [in 1912] to levy an income tax," he reminded his colleagues. "[But] conditions have changed, and we have the right to-day to exercise the privilege of levying these taxes wherever we think they will bear most lightly on the consumer." He added a comment on a longstanding difference between the two sides of the aisle: "The Republican Party is in favor of taxed wool; the Democratic Party has declared itself in favor of free wool."[53] Once again, partial funding of the federal government through a tax on the earnings of the well-to-do was directly compared with a tariff on the wool of foreign sheep—and this by the sponsor of the bill that launched the modern internal revenue system as we know it, one that pays exponentially more attention to paychecks than to fleeces.

———

As noted in chapter 3, wartime need for wool led to the United States having twice as many sheep at the end of the Civil War as at its commencement. But after the conflict, with the price of wool plummeting due to the flood of woolen military surplus garments, a large-scale slaughter of America's sheep took place. It was estimated that over four million sheep were destroyed in late 1868 and early 1869.[54] When the Civil War began, Ohio was by far the leading sheep state, with over 3.5 million head. By 1867 that number had swelled to over 7.6 million sheep—but then the bottom fell out, and Ohio registered fewer than 5 million sheep in 1870.[55]

The post–Civil War slaughter was not directly caused by tariff policy, and some of the declines in sheep numbers in Ohio and other states east of the Mississippi River were due to flocks being moved west and to losses from disease.[56] Nevertheless, the incident stuck in the minds of late nineteenth-century Republican lawmakers, especially in the Midwest, who made dire predictions that making wool duty-free would lead to more mass ovine killing.

In the 1894 debate, for example, Minnesota Republican congressman James T. McCleary warned that duty-free wool would lead to "the slaughter of flocks from ocean to ocean." Later in the same speech, he hammered the point even harder, arguing that the bill under consideration was "hostile legislation . . . threatening American sheep husbandry with complete extermination."[57] McCleary's colleague John H. Gear of Iowa agreed, declaring that having no tariff on wool would result in "the absolute annihilation of the American flocks."[58] Forty-seven million sheep would be sacrificed, Michigan's Julius C. Burrows predicted.[59] George W. Ray of New York made the same point as the others but used a vivid analogy. "[Pass this bill]," he said, "[and] American sheep will be as rare in a few years as American buffalo."[60] Clearly these men were far more concerned with the duty-free wool provision of the 1894 law than with its radical idea of taxing incomes in peacetime.

The debate over government revenue in the United States of our time has lost its former ovine focus. Arguments today over whether the income tax rates should be raised, lowered, or kept the same are not coupled with concurrent quarrels over whether the change will cause the nation's sheep farmers to destroy their flocks. Perhaps Americans dreading the annual ritual of reporting their earnings for the year to the Internal Revenue Service can at least take some comfort that when they submit their 1040 forms on April 15, this action, if not pleasant, at least will not be purported to lead masses of sheep to slaughter.

CHAPTER 9

"A Little Pure Wildness ... Both of Men and Sheep"

Big Flocks, Bighorns, and Western Conservation

> On the slow recession of those rock-grinding glaciers, at the close of the Glacial Period, this valley basin came to light: first a lake, then a sedgy meadow, then, after being filled in with flood and avalanche bowlders, and planted with trees and grasses, it became the Yosemite of to-day—a range for wild sheep and wild men.—JOHN MUIR, "The Wild Sheep of the Sierra"

In the 1980s Dr. William Foreyt of Washington State University conducted an experiment with gut-wrenching results. He and his colleagues placed a group of six sheep on a five-acre pasture. These animals—call them flock A—spent a year in the large enclosure grazing and living their lives normally. Then, six new sheep—call them flock B—were added to the pasture. The newcomers were healthy, or, as the veterinary literature put it, "clinically normal."

Four days after the introduction, one of the sheep from the first group was dead. Within another month, four more of the A flock had perished. The last sheep from group A finally succumbed seventy-one days after the newcomers were added. Meanwhile, all the sheep in the B flock were doing just fine.

What happened to the unfortunate sheep from A? Foreyt determined that they all died of acute hemorrhagic pneumonia, which they had contracted from the animals in the second group. But if that strain of pneumonia was so deadly, why didn't the sheep in group B also die since they carried the bacteria that caused the illness?

The answer is that the sheep were of two different species. The ones in group B were domestic sheep, *Ovis aries*. Group A's animals, on the other hand, were bighorn sheep, *Ovis canadensis*, a close relative to the domestic livestock, and one of two species of wild sheep native to North America (see fig. 9.1).[1]

The abstract of Dr. Foreyt's paper warns: "On the basis of results of this study and of other reports, domestic sheep and bighorn sheep should not be managed in proximity to each other because of the potential fatal consequences in bighorn sheep."[2] Foreyt's study was a controlled experiment using captive animals. But many of the populations of bighorns in the West are actively monitored, so, if his conclusions were right, it was likely that at some point his experiment would unintentionally be repeated in the field due to a domestic sheep wandering into the range of its closely scrutinized wild relatives.

This happened in 1997 when wildlife biologists studying bighorns living in the Tarryall and Kenosha Mountains of central Colorado suddenly began finding carcasses of wild sheep that had died of pneumonia. The biologists soon saw a single domestic sheep grazing near some bighorns; it was the first time they had seen a domestic sheep in the study area in six years of research. The sheep was shot, but the damage was done. The resulting epizootic—caused by contact with just one domestic animal—killed eighty-six bighorns. Nine years later the bighorn population still had not recovered; the Tarryall herd was only about half as large as it was in 1997.[3]

Of course, domestic sheep and bighorns had comingled in the West for well over a century before Foreyt and the other scientists published their articles. Armed with the knowledge gleaned by recent studies, some have argued that the long-term decline of bighorn in the West is not simply the result of habitat destruction but also a consequence of introduced livestock. This could explain some massive die-offs from the past. For example, at Pikes Peak in Colorado, a population of about three hundred bighorns in 1952 fell to fewer than forty animals by the following year. It seems likely that pneumonia contracted from domestic sheep was the cause.[4]

That vast numbers of people native to the New World died of smallpox and other diseases following contact with Europeans is well known, since it has been described in several books intended for general audiences.[5] That America's bighorn have suffered—and continue to be impacted—by proximity to domestic sheep of European descent is less well known, since it is mostly described in scientific literature.

Yet this susceptibility of bighorns to sheep-borne pathogens remains a prime concern and a management challenge for today's wildlife biologists. Small wonder that the Western Association of Fish and Wildlife Agencies, a consortium of twenty-three state and provincial wildlife agencies from

FIG. 9.1. A North American bighorn sheep ram. Library of Congress, reproduction number: LC-DIG-highsm-38962.

the western United States and Canada, formed the Wild Sheep Working Group to address the problem. In 2012 the group published a report titled *Recommendations for Domestic Sheep and Goat Management in Wild Sheep Habitat.* The report advocates that all parties involved in bighorn conservation maintain as their primary goal "effective separation"—which they define as "spatial or temporal separation between wild sheep and domestic sheep or goats to minimize the potential for association and the probability of transmission of diseases between species."[6] This sounds easier than it is; the twenty-four-page document acknowledges the difficulties faced in implementing the proposals. For example, it notes that in some cases a nine-mile buffer zone between domestic sheep and bighorns has been effective, but this assumes that a bighorn herd is reasonably sedentary in its home range. "In contiguous wild sheep habitat where movements by wild sheep have the potential to exceed *a priori* expectations, buffer zones may not be effective or practical," the report cautions.[7] It might have said the same thing about domestic sheep; if there is one thing these two genetically similar animals have in common, it is that neither confines its wanderings to suit the wishes of people.

Keeping wild and domestic sheep separate wasn't a concern for sheep advocates of late eighteenth-century America. There was no way it could have been; they did not know that wild sheep existed in the lands beyond the Mississippi River. Not until 1800 were specimens of the bighorn collected and brought east for scientists to examine and classify.[8] But even well after bighorns became familiar to nineteenth-century Americans, little attention was paid to the well-being of the wild sheep as the domestic flocks were driven west. It wasn't just domestic sheep's competition with bighorns or other wildlife that was becoming chronic, however. Some protested that simply by their grazing, the vast flocks of sheep were harming pristine western landscapes. Prominent among persons objecting to all those sheep was a Scottish immigrant recognized as a major figure in the conservation movement.

In a well-known image of John Muir, he stands beside Theodore Roosevelt at Glacier Point, high above Yosemite Valley in California (see fig. 9.2). The two men had a mutual love of the outdoors and a belief that America's spectacular lands should be preserved as refuges. They had another thing in common: their dislike of sheep. But whereas Roosevelt's ovine disgust

FIG. 9.2. Theodore Roosevelt and John Muir at Glacier Point, above Yosemite Valley, California, 1903. Both men disliked sheep, but no doubt they are wearing wool in this photo. Library of Congress, reproduction number: LC-DIG-ppmsca-36413.

centered on sheepmen getting in the way of his cattle ranching, Muir's aversion arose because he himself had been one of those shepherds Roosevelt had called "a morose, melancholy set of men."[9] Muir's distaste for sheep, then, was partially grounded in personal experience with the animals. Also, unlike Roosevelt, Muir sometimes expressed pity for sheep rather than hostility. But the most significant aspect of Muir's attitude toward sheep was how it shaped his advocacy for conservation.

Muir was born in Scotland in 1838, and when he was eleven his family immigrated to the United States, settling on a farm in Wisconsin. The young Muir was quite transient; eventually he arrived in California in 1868, beginning the phase of his life for which he is best remembered today.[10] Needing employment, Muir accepted an offer of a dollar a day plus board to look after a flock of eighteen hundred domestic sheep in the Twenty Hill Hollow area of Merced County. Muir enjoyed being outdoors, and he reflected in his journal on the beauty of nature, but he didn't care for being a shepherd. One day he heard frightened bleating; looking in the direction of its origin, he saw that two coyotes had killed one of his lambs. What followed in his journal was not the venom toward coyotes that a typical sheepman might have expressed but rather a respectful appreciation of the wild canines. "They are beautiful animals," he wrote, "and although cursed of man, are loved of God." He added: "Their sole fault is that they are fond of mutton." Muir then contrasted the predators with his charges. "The sheep of my flock are unhappy creatures, dirty and wretched, miserably misshapen and misbegotten, and I am hardly sorry to see them eaten by those superior beings."[11]

Despite his disagreeable experience, Muir took a second, similar position in July 1869, helping drive over two thousand sheep to the headwaters of the Merced and Tuolumne Rivers, in the Sierra Mountains just north of Yosemite Valley.[12] Muir's appreciation of the region's natural beauty increased. So did his abhorrence for sheep, and he was delighted when his shepherding days ended. But forever after he was alert to the damage sheep caused. In an article he wrote in 1879 for *Scribner's Monthly*, Muir described his alarm at seeing human tracks around Shadow Lake, one of his favorite secluded spots in the Sierras. He surmised that the tracks were made by a shepherd, adding that his worst fears were realized when he discovered more than just tracks: "A trail had been cut down the mountain-side from the north, and all the gardens and meadows were destroyed by a hoard [*sic*] of [hooved] locusts, as if swept by a fire. The money-changers were in the temple."[13] Decades later Muir wrote an article for the *Atlantic Monthly* on his first summer in the Sierras. Extolling the beauties of a mountain meadow, he declared, "To let sheep trample so divinely fine a place seems barbarous."[14] As founder of the Sierra Club, he made it his life work to insure that this wouldn't happen.[15]

Like his friend Theodore Roosevelt, John Muir contrasted the ugliness he saw in domestic sheep with the magnificence he beheld in their wilderness cousin, the bighorn. In an 1874 article titled "The Wild Sheep of California," Muir wrote: "In form the domestic sheep is expressionless, like a round bundle of something only half alive; the wild is elegant as a deer and every muscle glows with life. The tame is timid; the wild is bold. The tame is always ruffled and soiled; the wild is trim and clean as the flowers of its pasture."[16]

Muir elaborated on this theme in his essay "Wild Wool" published the following year. He told of his careful examination of the fleeces of three bighorn sheep. Concluding that their wool was superior to that of the domestic animal, he was moved to exclaim: "Well done for wildness! Wild wool is finer than tame!"[17] His friends, however, were not impressed by this observation; they objected that while the mountain sheep may have had finer wool, each animal had too little of it for practical use. "How many wild sheep . . . would be required to furnish wool sufficient for a pair of socks?" they demanded. Ever the defender of the wilderness, Muir retorted that they were missing the point. "Wild wool was not made for men but for sheep," he asserted, adding, "However deficient as clothing for other animals, it is just the thing for the brave mountain-dweller that wears it."[18]

To this point the article seems the standard Muir exposition on nature, but Muir took a surprising turn. He lamented that in spite of thousands of years of selective breeding of sheep, little progress had been made: "We still seem to be as far from definite and satisfactory breeds as we ever were. In one the wool is apt to wither and crinkle like hay on a sun-beaten hill-side. In another, it is lodged and matted together like the lush tangled grass of a manured meadow. In one the staple is deficient in length, in another in fineness; while in all there is a constant tendency toward disease."[19]

Muir then offered his startling prescription for sheep deficiencies, connecting the prescription to his acclaim for the fineness of wild sheep fleece. Why not, Muir asked, round up several hundred of those majestic wild sheep and breed them with their lowly domestic cousins? "I am ready to undertake their capture," he added helpfully.[20] In an America that had debated and practiced sheep improvement for over a century, this was the most radical of all proposals; it would in part be taking domestication back to its roots. Muir was confident it would work. As he put it in the final sentence of the essay, "A little pure wildness is the one great present want, both of men and sheep."[21]

⸺∽∽∽⸺

Had large numbers of bighorns actually been caught and put with domestic sheep, the result would have been what William Foreyt observed over a hundred years later: most or all of the wild animals would have died suddenly, and the nineteenth-century would-be improvers would have been astonished by this. Perhaps if Muir in "Wild Wool" had applied a bit more of his typical suspicion that humans can improve things, he would have instead endorsed "effective separation" to protect the bighorn, long before the 2012 proposal of this concept by the Wild Sheep Working Group.

His surprising musings on comingling domestic and wild sheep aside, Muir desired that certain unspoiled places be forever protected from all human interference. In the Sierra lands he revered, he got this wish. The first legislation to preserve the area's natural character was enacted in 1864, several years before Muir arrived in the area. Congress that year granted Yosemite Valley and the Mariposa Big Tree Grove to the State of California to hold "for public use, resort, and recreation."[22] This official designation, however, was insufficient to prevent continued private abuse of Yosemite's resources, including grazing by sheep.[23] Muir and others campaigned for more stringent protection, advocating for expansion of the protected areas and for federal rather than state oversight. These efforts were successful: the genesis of what would become Yosemite National Park was an 1890 act of Congress setting aside additional lands as "forest reservations" under the auspices of the federal Department of the Interior. The legislation further specified: "[The reserved lands are] hereby reserved and withdrawn from settlement, occupancy, or sale under the laws of the United States . . . and all persons who shall locate, or settle upon, or occupy [the designated areas] shall be considered trespassers and removed therefrom."[24]

All this was an expansion of government authority that, like other conservation efforts of the late nineteenth and early twentieth centuries, required a rethinking of constitutional limits. The nation had, after all, been founded on the principles of limited government, of powers reserved to the individual states, and of private property rights. Objections were raised that by designating reserves and thereby removing land from settlement, Congress was violating all of these founding principles.[25] Given the somewhat rival concepts of property rights versus preservation, it was inevitable that Alexis de Tocqueville's observation that "there is virtually no political question in the United States that does not sooner or later resolve itself into a judicial question," would soon play out, with shepherds among the plaintiffs challenging federal authority.[26]

Litigation arose from laws expanding federal power to regulate human activity on protected lands. The Forest Reserve Act of 1891 gave the president of the United States the authority "to ... set apart and reserve ... forests ... as public reservations."[27] President Benjamin Harrison signed the bill, and during the remainder of his term he proclaimed eleven forest reserves, including one in the Sierras abutting Yosemite National Park.[28] (Under its current name, the Sierra National Forest, the reserve includes a section called the John Muir Wilderness.)[29] Later in 1897, Congress acted to further protect the designated lands by giving the secretary of the interior power to "make such rules and regulations ... as will insure the objects of such reservations, namely, to regulate their occupancy and use and to preserve the forests thereon from destruction."[30] This was followed by a provision giving the federal government the power to bring criminal charges against violators of those rules and regulations.

Preserving forests from destruction meant restricting or prohibiting grazing by sheep, and this did not sit well with the men John Muir accused of driving their flocks of hooved locusts through his beloved Sierras. Furthermore, when under the 1897 law a California man was prosecuted for grazing his sheep in a reserve, the federal government lost. In the 1900 case of *United States v. Blasingame*, federal judge Olin Wellborn dismissed the charges, holding that by giving the secretary of the interior authority to declare violations of the rules and regulations a crime, Congress had unconstitutionally delegated legislative power to an administrative agency.[31]

The Wool Growers' Association of San Francisco gleefully publicized the *Blasingame* decision among the state's sheepmen, assuring them that the ruling meant they could not be prevented from grazing their stock in the restricted federal lands.[32] The next year, an official reported that at least twenty-three sheepmen had herded about a hundred thousand sheep across the Sierra Forest Reserve.[33]

The federal government persevered. A reorganization in 1905 transferred the forest reserves from the Department of the Interior to the Department of Agriculture.[34] The following year, the secretary of agriculture issued a regulation prohibiting livestock owners from grazing their animals on a forest reserve unless they first secured a permit.[35] Finally, in 1907 Congress expanded the jurisdiction of the Supreme Court—when *Blasingame* was decided, the federal government could not appeal a lower court judgment in a criminal case, preventing the *Blasingame* prosecutors from presenting the matter to the highest court in the land. The so-called Criminal Appeals Act changed this.[36]

The *Blasingame* decision had been problematic, not just in its setback of forest preservation efforts but also because, coupled with other litigation at the time, it caused jurisprudential inconsistency. In some similar cases involving a federal prosecution, the defendant prevailed, as in *Blasingame*. In other cases, however, the ruling was in favor of the government.[37] Given this division in lower court rulings, it seemed certain the matter would have to be settled by the Supreme Court.

On a spring day in 1907, Pierre Grimaud and J. P. Carajous ignored the regulations and without a permit moved their sheep onto the Sierra Forest Reserve to graze. They were apprehended and charged. Their attorneys cited *Blasingame*, repeating the argument that, by giving the secretary of agriculture authority to issue rules and regulations, Congress was unconstitutionally delegating its legislative powers to an executive department.[38]

Grimaud prevailed in lower court, but the government won on appeal to the Supreme Court. In *United States v. Grimaud,* Justice Joseph Lamar wrote the unanimous opinion. He agreed that a long history of decisions had established as settled law that Congress alone possessed legislative powers; these they could not grant to any other part of the government.[39] The difficulty, he noted, came in trying to draw a line between impermissible legislative delegations and permitted administrative authority. In the matter of the sheep grazing permits, Lamar reasoned, the secretary of agriculture had acted within his proper scope. Where and whether such permits were necessary "was a matter of administrative detail," he wrote. The harm done by grazing sheep might differ from one forest to another—or even differ within a particular forest depending on the season of the year or the stage of timber growth. Accordingly, it would be "impracticable for Congress to provide general regulations for these various and varying details of management." Since conditions in each reservation were unique, "in authorizing the Secretary of Agriculture to meet these local conditions, Congress was merely conferring administrative functions upon an agent, and not delegating to him legislative power."[40]

Legal scholars have pointed out that the significance of *Grimaud* extends far beyond the immediate question of whether the Department of Agriculture could enforce a permit system to manage sheep grazing on a designated reserve. *Grimaud* subsequently operated as a broad grant of power to Congress to delegate regulatory authority to any federal agency, for any legitimate means.[41]

But from the standpoint of those working to save the forests, the beauty of *Grimaud* was its importance to their preservation efforts. In 1912, the year after the case was decided, there were forty-nine prosecutions for grazing trespass on forest reserves. Apparently the message from this crackdown got through, because by the end of the year the Forest Service reported that overgrazing was no longer a problem.[42] Muir and other conservationists had reason to celebrate. A nation that in its infancy had viewed the rapid expansion of America's sheep flocks as a desirable goal—even a critical one—in its second century was coming to understand that saving natural areas was also a prudent policy, one that needed to be undertaken even if it meant excluding those same valuable sheep from large portions of the West. The conservation movement that began in the late nineteenth century led to the establishment of wildlife sanctuaries—including ones providing that critical spacing needed so that America's wild bighorn sheep would not perish from diseases transmitted by their immigrant domesticated cousins.

CHAPTER 10

Swimsuits, Soldiers, and Polyester
The Decline of America's Sheep

> Modern military forces must have millions of specially designed uniforms and our military leaders insisted that many of these be made entirely of wool.
> —JOHN W. KLEIN, *Wool during World War II*

Some things don't change. The beach was a popular destination for vacationing Americans in the late nineteenth century, just as it is today. What has changed over the years—strikingly—is what people wear on the nation's beaches, not only the styles but also the fabric.

An article in the *Boston Weekly Globe* in August 1890, written by a woman, advised female readers on what to wear at the beach. Part of the guidance offered was economic. The article noted that many women were distressed by the high cost of store-bought swimwear. This was not an insurmountable problem, the budget-minded consumer was assured, because it was a simple matter to convert an old dress into a delightful bathing suit (see fig. 10.1).[1]

Once she decided to take this on as a do-it-yourself project, a woman would have to take care as to the material of the garment. While her bathing suit should be as thin and light as possible, the article cautioned, "It should not cling to the figure . . . when wet." And it was for this reason, the reader was solemnly told, that "cotton goods should never be used." Better that the suit be composed of a durable wool fabric—or, if not that, how about cashmere or alpaca?[2] All of the options proposed, in other words, were the product of sheep or other fiber-bearing animals.

Wool bathing suits were not strictly Victorian attire. Wool remained prominent in beachwear well after World War II. An advertisement from the April 1, 1949 issue of *Vogue*, promoting a "Wool Modern" swimsuit:

FIG. 10.1. Late nineteenth-century women's wool bathing suits, *Boston Weekly Globe*, August 2, 1890.

"Modern—because this swimsuit is made of featherlight zephyr wool with Lastex control (see fig. 10.2)." The "bare-shouldered swim sheath" was "created by Cole of California to fit and flatter lovely contours" and was said to be available "at the better stores, [for] about twenty dollars."[3] Seeing the contrast with figure 10.1, it is evident that styles women sported on the beach had become more revealing in six decades, but sheep provided the material for the swimsuit in both.

Today there is a tendency to view the uses of wool very narrowly. Adults with young children observing the sheep at my museum have a tendency to say, "They cut the wool off the sheep and use it to make sweaters!" Sweaters, the children already know, are worn when it is chilly, and thus the perception of wool as used only for one particular article of cold-weather clothing attaches to the young mind. As we have seen throughout this book, however, in its heyday, wool was the fabric of choice for many garments—everything

FIG. 10.2. Women's wool bathing suit, *Vogue*, April 1, 1949.

from winter coats to bathing suits. But in the decade following the 1949 *Vogue* ad, wool's ubiquity would decline, and with the decline began a reduction in America's flocks of sheep. This chapter examines how and why this happened. It is more a story of clothing worn on battlefields than of clothing worn on beaches.

It had only been two months since the attack on Pearl Harbor brought the United States into World War II. Americans understandably looked for reasons to be optimistic that their nation and its allies would prevail in the massive global conflict it had been drawn into. Readers of the *Los Angeles Times* encountered an article assuring them that all would be well. A key factor would enable victory over Hitler and the other totalitarians: the free world had most of the sheep.[4] Noting the recent setbacks of the German army in Russia, the article pointed out that the invading troops were ill prepared for the severe winter they were experiencing. Small wonder, the writer of the piece elaborated. Hitler had about six million soldiers in the field, and Germany thus needed eight hundred million pounds of wool annually—three to four times the yearly wool clip in all of Europe (see fig. 10.3).[5]

The United States and its allies, the article insisted, were in a much better position. Half the wool harvested in the world came from Australia, New Zealand, Tasmania, and South Africa, while the United States contributed another 450 million pounds to the West's annual clipping. Altogether, the Allies harvested sixteen times more wool yearly than the Axis. American soldiers would not suffer the shortage of warm clothing plaguing the Nazis, readers were assured. "Fortunately for the democracies," the article concluded, "the sheep of the world seem to be working against Hitler."[6]

Given the tenor of the times, it might have been unpatriotic to point out that, as noted in chapter 1, the United States had prevailed in its war of independence against Great Britain in spite of *not* having a decided wool advantage. In war, the side with the most sheep wasn't guaranteed victory. Having superior stocks of wool was a good start, but much more would be needed to win World War II.

In those anxious first months of 1942, the *Los Angeles Times* could be optimistic. But in Washington the government was nervous as it considered the staggering amount of wool that would be necessary for the United States

Swimsuits, Soldiers, Polyester

FIG. 10.3. On February 8, 1942, a story in the *Los Angeles Times* included an illustration similar to this one, highlighting the advantage the United States and its Allies had over the Axis powers in terms of available wool. Above this newspaper graphic was a caption reading "While Nazi troops freeze and Hitler shrieks for warm clothing, millions of sheep are growing wool that will carry the Allies to victory." Illustration by Selena Lim.

to win the war. In the years of peace prior to Pearl Harbor, the annual consumption of wool was only 2.2 pounds per capita. But war was a different matter. Estimates—made not by the military but instead by the U.S. Tariff Commission—were that to supply the initial issue for each soldier in the army, seventy-five pounds of cleaned wool were needed. This was just for soldiers' training. For combat the figure would rise to one hundred pounds per soldier in the first year. Annual maintenance thereafter would be much less, only forty pounds per soldier a year. Still, with an army of 3.75 million men forecast for 1942 and 2 million additional men in 1942, the necessary amount of wool was astounding.[7] The United States, if these predictions held, would use up to a billion pounds of raw wool annually during the conflict. That would prove to be an underestimate.[8]

Where was this huge quantity of wool to come from, given that America at the time was only producing around 450 million pounds a year, far short of what was needed? Clearly, foreign supplies would be required. Medium grades of wool could be procured from South America, but these were suitable mainly for blankets and overcoats. Also needed annually, according to forecasts, were at least 200 million pounds of imported, fine, apparel-grade wool. This could only be procured from Australia and South Africa.[9]

From America's view of the world in 1942, this was a big problem. The Japanese had taken control of most of southeastern Asia and seemed poised to invade Australia. While an invasion of South Africa was less of an imminent threat, an interruption to shipping around the Cape of Good Hope was a possibility, given the German positions on the continent.[10] The United States could not count on these Southern Hemisphere wools. What was to be done?

A Tariff Commission employee had an imaginative plan to deal with the shortage. He envisioned increasing the wool supply with the help of America's rural youth.

∽∂∬∂∽

A Tariff Commission 1942 report detailed some of the fine points of sheep husbandry. Dog-proof shelter at night for sheep was a must, it advised. A flock of twenty cross-bred ewes of average size would need four to five tons of hay to sustain it through the winter, supplemented by about five hundred pounds of "bran, oats, or similar feed" so that they would produce healthy lambs. Of course, in milder climates with a longer grazing season, less winter hay would be necessary. These words could have come from any of countless books or agricultural journal articles published at any point in American history. One might guess that the author, Louis G. Connor, was a nineteenth-century gentleman farmer or perhaps a late twentieth-century extension service agent.

In fact, Connor bore the austere title of "principal commodity specialist" for the Tariff Commission. His comments on proper care of sheep appeared in his March 1942 report titled "A Tentative Plan for an Emergency Increase in Wool Production in the United States."[11] Connor began the report with a solemn note that the threat to shipping routes and the massive amount of foreign wool needed for military gear meant that a wool shortage might soon develop.[12] He offered a proposal to combat this possible crisis. Sheep ranchers in the western states, Connor noted, normally culled older ewes, generally after they had given birth to five lambs. This was a logical policy, since resources are finite even on an open range. Why keep a four- or five-year-old ewe when it could be slaughtered and its place in the grazing flock occupied by a younger, more vigorous animal? Connor calculated that there were nearly four million "aged or near-aged" ewes scheduled for such culling over following next two years, but he pointed out that these older ewes generally still had one or two years of potential reproductive ability. Accordingly, he suggested that instead of culling, these ewes be bred. Following this, flocks of twenty ewes each would be sent to be taken care of by teenaged boys and girls living on farms, youths who were likely members of local 4-H clubs. The government would provide the rams to breed with the aged ewes in successive years. Under Connor's plan, the young farmers would assume all responsibility for the care of the ewes, but these animals would remain property of the government, which would collect all the wool sheared from them. Lambs born to these old ewes, however, would become property of the 4-H youths. By agreement, the lambs would be slaughtered soon after weaning ("to insure that these Government sponsored flocks retain their emergency character"). The government would purchase the slaughtered lambs to turn into rations for American troops. By way of such a plan, Connor asserted, the young shepherds would gain "a valuable, and, for them, profitable experience."[13]

The contrast is striking between Connor's plan and the Chicago sheep club described in chapter 6. In Chicago, faced with a wool shortage during the First World War, the notion was promoted that even city dwellers could help by keeping sheep on their lawns. Connor, on the other hand, proposed a purely rural effort. In the quarter-century since the U.S. entry into World War I, urban growth had continued. While sheep on urban lawns in 1917 might have seemed quaint but not totally out of place, such a notion had grown incongruous twenty-five years later.

As it turned out, Connor's proposal was never implemented. A postwar critique of his idea pointed out that it would have raised the domestic wool supply only minimally—only 6 percent from 1943 to 1945, with most of this modest increase not emerging until 1944.[14] If the American cause in World War II had truly become dire and every little bit of additional wool was precious, perhaps Connor's suggestion would have been put into practice. Fortunately, there was never reason to commit money and resources to a government-run program to expand the nation's flock of sheep. This was possible because of a setback in Axis military fortunes and sound planning by the Allies before the war and in its early stages.

Axis invasions of Australia or South Africa never occurred. Allied advances in the South Pacific were so successful that by early 1943 the US

Department of Agriculture suggested that further importations of wool be halted and even that existing stocks be liquidated to avoid a negative economic effect on American woolgrowers and domestic markets. The Army Quartermaster Corps was initially opposed to this, but continued American military triumphs led the government to begin auctioning its glut of wool by January 1944.[15]

That the feared World War II wool crisis did not occur was also attributable to actions taken by the US government and by the private sector before the United States was drawn into the conflict. On the private side, by December 1940 U.S. importers began bringing into the country approximately 41 million pounds of apparel-grade wool a month—about six times the amount imported in the prewar years.[16]

Also crucial to Allied wool fortunes was cooperation between the United Kingdom and the United States. Within weeks of Hitler invading Poland, the British government took control of all wool in Australia and New Zealand, and in August the following year they instituted the same policy in South Africa. Essentially, sheep in these British Dominions became four-legged employees of the Crown. This put three-fourths of all apparel wool grown outside the United States under the authority of the United Kingdom.[17] As noted, it was feared that the countries providing this abundance of sheep were possible targets for Axis attack. Furthermore, the United States would rely on that wool just as the British would. Thus, the two countries entered into a series of agreements, under which a reserve of Southern Hemisphere wool would be stored in the United States, and both the United States and Britain could draw from this reserve in an emergency.[18] Wool from Australia and New Zealand was transported to the United States on supply ships returning home from the South Pacific after supplying American forces with munitions.[19]

In the early stages of the war the U.S. government considered exercising total control of American wool, from sheep to factory to retailer, and also rationing civilian clothing, but the U.S.-British measures worked so well that this wasn't necessary.[20] The world's sheep had indeed served the Allies well in their campaign against totalitarianism, just as the *Los Angeles Times* had predicted.

But peace would be fleeting. Post–World War II liquidation of surplus wool would come back to trouble the United States five years later when it faced a new military challenge.

Senator Lyndon Johnson of Texas was irate. "Obviously we are badly in need of more courage and more initiative in some high places," he declared in December 1950. "The country would be a lot better off if our defense officials and others would tell us bluntly what the country needs rather than what they think the country needs to hear."[21] Johnson was chair of the Senate Armed Services Preparedness Subcommittee. His angry call for more courage and initiative accompanied a blistering report he and the other senators on the subcommittee issued criticizing the Army and Navy Munitions Board of the Department of Defense for its failure to stockpile enough wool to meet the sudden emergency caused by the outbreak of the Korean War earlier that year.[22]

Johnson's subcommittee was particularly vexed given recent history demonstrating America's need for wool in times of international conflict: "The Munitions Board . . . has chosen to ignore not only the hard-learned lessons of World War I but of World War II also. In both wars, we were short of adequate wool supplies. . . . If general mobilization were undertaken now, we would again be as bad off—or perhaps even worse—[than] we were during both World Wars."[23] The report noted that the United States only produced about one-fourth of the wool it consumed, meaning a huge dependence on foreign sources. This would cause a particular problem in the Cold War: "Foreign wool is not ours merely for the asking; we have to buy it on the open market, and, needless to say, other nations—including nations behind the iron curtain—are bidding for the available supply."[24] Potential enemies could thus stand between America and its military need for wool.

Why had this unfortunate situation come about? The Preparedness Subcommittee blamed bureaucratic bungling. It noted that as early as 1947 the secretary of agriculture had written to the Munitions Board noting that the Commodity Credit Corporation (CCC)—a government-owned corporation housed in the Department of Agriculture—had about 460 million pounds of wool, much of it suitable for military fabrics. The Preparedness Subcommittee urged the board to transfer the wool to its stockpiles.[25] But due to its standing as a corporation, the CCC could not simply hand over title to the wool without reimbursement. The Munitions Board, not having the authority to buy it, said it would accept the wool if it was given but could not purchase it. Thus there was an impasse between two units of the government. The CCC pushed for Congress to change the law so that the transfer could go through, but no such bill was forthcoming. Instead, Congress authorized the CCC to begin liquidating its wool stocks, which it did, and by June 30, 1950—five days after the start of the Korean War—the liquidation was complete.[26] As a result of this, the subcommittee complained, "Today the United States—faced with the prospect of a long winter campaign in Korea and the maintenance of a 3,000,000-man armed force—has no wool in a stockpile, no wool in inventory, and less than enough wool in prospect through our domestic production."[27]

The press covered this matter as a scandal. Shortly after issuance of the Preparedness Subcommittee's report, an angry newspaper columnist wrote: "If some GI finds himself off fighting the Reds in 20-below temperatures, thinly clad in a pair of cotton pants, he can blame one of the biggest red-tape, buck-passing snafus ever to hit the capital."[28]

The Munitions Board argued that it was not at fault for the failure to stockpile wool because the supply offered by the CCC was too expensive. The price quoted was two and a half times larger than the board's appropriation for stockpiling *all* supplies. Furthermore, the board noted that experts in 1947 said there would be no wool deficit in the foreseeable future, even that there was a surplus that would take thirteen years for the market to absorb. Finally, there was the cost of rotating and maintaining such a quantity of wool to consider—this would amount to 10–15 percent of the value of the wool every five years. For these reasons, the Board concluded, "A dollar invested by the Government in the stockpiling of wool will yield much less protection than the same dollar invested in 71 other materials being stockpiled."[29]

It's hard to miss the similarities between Lyndon Johnson's angry comments about the Munitions Board in 1950 and George Washington's exasperated letter to the Board of War in 1778, saying that the supply of blankets was insufficient. In both cases, during a time of war a future U.S. president was commenting on a military need for wool that was unfulfilled. Washington and Johnson both recognized that the animal so often viewed as an emblem of peace and tranquility provided an essential commodity for troops in battle.

The world Lyndon Johnson inhabited at the dawn of the Korean War was vastly different from the one that George Washington experienced during the American Revolution. Washington traveled by horse, but when Johnson served in the Senate there were planes, trains, and automobiles. In Johnson's America of 1950, radio and telephones were well established, while television and computers were being developed. Washington couldn't have dreamed of such things. Weapons of war had grown ever more powerful (and their results more gruesome). Johnson was sworn in as U.S. senator less than four years after atomic bombs were dropped on Hiroshima and Nagasaki—a far cry from General Washington's muskets. But in spite of all the technological advances made between 1778 and 1950, American servicemen serving in the war in Korea were hampered by the same difficulty that their ancestors faced fighting at Valley Forge—a shortage of wool. Combat depended on the work of soldiers, but much of their clothing still relied on the labor of shepherds.

But a transformative innovation was about to take place.

In response to Johnson's subcommittee's accusations, the Munitions Board never denied that there was a shortage of wool. They argued, however, that this deficiency could be compensated for in several ways. Domestic production could be increased, and civilian domestic consumption could be reduced. Imports could be expanded. And, as a last resort, the board maintained, substitutes for wool could be used.[30] A prominent American manufacturing company had been working for years on just such a substitute.

No, no, no, William Hale Charch insisted to his colleagues at the DuPont Company. The synthetic fiber we develop does not need to have the same physical structure as wool. It only needs to have the essential *properties* of wool. Charch, hired in 1924 as a research manager, would remain with DuPont until his death in 1958, rising to the position of director of pioneering research in the Textile Fibers Department.[31]

DuPont researchers focusing on wool's structure were inspired by their microscopic examination, which revealed that each strand was covered with serrations resembling scales (see fig. 10.4). This characteristic surface of wool enables the strands to readily entwine together, giving fabric made from wool its resiliency.[32] An image of Merino wool fiber taken through an electron microscope clearly shows these tiny serrations.

From the days of antiquity when humans first wove cloth until the twentieth century, the options for textiles were limited to biological sources.

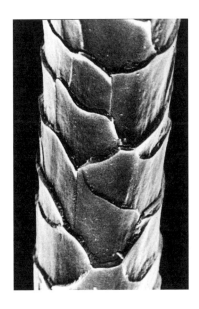

FIG. 10.4. Electron microscopic image of a Merino sheep wool fiber. Commonwealth Scientific and Industrial Research Organization (Australia).

These could be fibers from plants such as cotton, linen, or hemp. From the animal kingdom there was silk from the cocoons of silkworm moths. And there was wool, coming primarily from sheep but also from a few other mammals, such as llamas, alpacas, and certain breeds of goats.

Human dependence on natural fibers began to diminish in the 1930s when DuPont developed Fiber 66, the first synthetic substance capable of being processed into filaments and yarn. Eventually they settled on the name "nylon" for this product. DuPont did not create Fiber 66 merely as a curiosity. The company took note that every year about $70 million worth of silk went into the production of silk stockings, with the average American woman purchasing eight pairs per year.[33] Thus, if DuPont could fashion the nylon into an acceptable substitute, they had a ready market for the product. Nylon first went on sale in the spring of 1940 and was an immediate sensation.[34]

At the same time as the nylon research, a few of DuPont's inventers were also engaged in developing what they called Fiber W. While nylon was envisioned as a substitute for silk, Fiber W was envisioned as a replacement for wool. Initially this project was a small diversion pushed to one side as the company focused on nylon. This was an economic decision, based on biology. Sheep being more copious producers of fleece than silkworm moths

FIGS. 10.5 & 10.6. The author's coat, vest, and jackets. All these garments are made not of wool but rather of a wool substitute, polyester.

are of cocoons, the price of wool was far less than that of silk. This made DuPont doubt it could develop a cost effective wool substitute.³⁵

That would soon change, however. Building on Charch's assertion that a synthetic fiber only needed wool's resiliency, not its characteristic structure, DuPont chemists eventually refocused their efforts and developed Fiber V, or, as we now know it, polyester. DuPont began selling the new product under the trade name Dacron in 1951.³⁶ This was exactly 150 years after E. I. du Pont crossed the Atlantic Ocean with Don Pedro, the Merino ram that was a first step toward American sheep improvement. A decade after Don Pedro's journey, du Pont opened a woolen mill near Wilmington, Delaware, one of the earliest such operations in the country. In the mid-twentieth century, the company founded by du Pont—a man intimately connected with the American goal of establishing better flocks of sheep and turning their wool into domestically manufactured textiles—was at the forefront of developing and marketing a workable artificial substitute for wool. A nation that had long fretted about the number and condition of its sheep, historically frustrated by fluctuating wool prices and with dependence on foreign flocks, and struggling in the 1950s to outfit soldiers fighting in the cold Korean winters, took note of the DuPont Company's invention.

The success of the artificial fiber developed by William Charch's team is apparent today to anyone in a temperate climate shopping for winter clothing. Coats and jackets that once would have depended on wool for their warmth are now insulated with polyester (see figs. 10.5 and 10.6). Nor is polyester a mainstay only in outerwear. It is used in many types of clothing, sometimes by itself, sometimes blended with other fabrics. Men's dress shirts, for example, commonly are a cotton-polyester blend, the cotton providing softness and comfort, the polyester serving to prevent wrinkles.[37] Gone is the wool attire sported on American beaches before the mid-twentieth century; synthetic fabrics are now almost universal in bathing suits. Sheep wool has been supplanted by chemical formulas.

༺༻

A 2004 publication by the U.S. Department of Agriculture spelled out how precipitous the depletion of America's flocks had been since World War II. In 1945 the nation had 56 million sheep, an all-time high. But by the early years of the twenty-first century, the United States was down to fewer than seven million.[38] This steep reduction has naturally led to commentary on its causes. For example, a 2013 article is titled "The Long, Slow Decline of the U.S. Sheep Industry."[39] Its author notes that synthetic fibers have reduced the demand for wool. The article also mentions diminishing consumption of lamb among Americans—although, as noted in chapter 7, lamb and mutton have never enjoyed the level of popularity in the United States as in some other nations, such as the United Kingdom and Ireland. The fondness of people in Europe for sheep meat initially led to an increase in American flocks during the late nineteenth century, when ovines on the range in Wyoming and Montana were shipped by rail to Chicago, slaughtered, and then sent overseas to hungry consumers. Clearly the story of the decline of sheep in the United States is more complex than can be explained by the abundance of polyester jackets and the dearth of lamb dinners.

A significant factor leading to a decrease in sheep was an increase in dairy farms, facilitated by urban growth. In 1919, the US Tariff Commission held hearings on the state of wool in the nation. A Michigan sheepman testified, "In localities especially favored with transportation to the large centers of population, dairy cows have driven out the sheep."[40] This, he asserted, was what had happened in Livingston County in his state—sheep had been common ten years earlier, but the county was close to the sprawling city of

Detroit, providing a nearby market for milk, more profitable than wool. So, out went the sheep and in came Holstein cattle. People from other states echoed his remarks.[41]

Dairy benefited not only from the rise of cities but also from the development of refrigeration—the same technological advancement that had made it possible for those sheep raised and butchered in America to be served safely at tables across the Atlantic Ocean. Raising sheep for wool and raising cattle for milk have several things in common. In both cases, the farmer depends on keeping significant numbers of a large, ruminant mammal so that their produce may be harvested. These sizeable flocks or herds require substantial resources to keep them in condition to yield the wool or the milk, especially in regions where the grazing season is limited so that livestock must be heavily supplemented to get them through the winter in good health. Also in common, both wool and dairy are renewable resources: unlike raising livestock for meat, the death of the animal is not required for harvest. One does not slaughter a sheep for its fleece or a cow for its milk. This means that each individual animal can provide a livelihood to its owner for an extended period of time, a valuable economic characteristic.

But there is one huge difference between these two forms of animal agribusiness: milk spoils; wool does not. The woolgrower need not get his product to market rapidly; well-packed fleeces can be stored without climate control in a barn or a warehouse awaiting transportation to factories days, weeks, or months in the future. Milk, on the other hand, sours quickly once it has left the cow's udder. If it is to remain a viable food for humans, steps must be taken to retard spoilage. One way to do this is through making the milk into cheese, a process humans have known for at least seventy-five hundred years.[42] But for the vast majority of human history, raising cows for fresh milk was simply not economically feasible. Refrigeration changed this. The possibilities of dairy expanded dramatically, and raising dairy cattle grew as an alternative to raising sheep.[43]

Those testifying at those 1919 Tariff Commission hearings on the switch from sheep to dairy cows were doing so at a time when the need for wool had neared its apex. The census the following year would show that America's sheep population was over one hundred million.[44] Synthetic substitutes for wool still lay three decades in the future, and the age of humans facing the cold in animal skins or furs had long past, along with much of the wildlife that had provided these.[45] Accordingly, if Americans in temperate climates in the 1920s were to get through winters without "starving with cold" as Reverend Seabury had put it a century and a half earlier, a reliable supply of wool was essential.

Yet even in those years before polyester, some sheep farmers began to go into dairy instead. It is a reminder that when production of one agricultural commodity declines, the answer to why this is occurring might not be that demand for the commodity is waning but rather that another crop, or another type of livestock, is simply a better bet for the farmer. It is also a reminder that to fully understand the reasons why America's flocks of sheep diminished in the second half of the twentieth century, we need to consider economic factors that existed much earlier. We saw that in spite of a substantial demand for coarse wool to clothe enslaved people during the antebellum years, American sheep farmers shunned this ready market to concentrate on raising sheep with higher-quality wool. They did so to try to maximize their profits. For the same reason, some sheep farmers generations later would move to raising dairy cows. Agriculture is a tough business, and those working in it can always be expected to seek new opportunities for greater fiscal security.

"Is that a *sheep*?" the astonished middle-aged woman asks me. I have placed a halter on Daisy, one of the Atlanta History Center's very friendly ewes, so that I can take her out of the barnyard and walk her around the campus. I respond that yes, Daisy is a sheep, and the woman exclaims: "I've never seen one of those before!"

The visitor's surprise causes me to think of Daniel Lee, so displeased that he saw only one sheep on a weeklong journey through Georgia in 1847. There were other parts of the United States at the time in which he would have seen a plethora of ovines. Indeed, as already noted, in 1840 the United States had more sheep than people. But today the notion of an adult American purporting to see a sheep for the first time is not outlandish. Even if a motorist in rural America eschews interstates and drives along two-lane highways, the agricultural landscapes are dotted frequently with cattle, occasionally with horses, but very, very seldom by sheep. No wonder an old West with far more sheep than cattle seems so incongruous today. Not only do television programs and motion pictures set in the nineteenth-century

West depict America as a country where cattle dominate, the experiences of twenty-first-century Americans reinforce that perception.

We have seen time and again in this volume, however, how significant sheep are to American history. The United States had to overcome a shortage of sheep to prevail in the Revolutionary War. The need for wool made sheep integral to American efforts in later wars as well. As the country moved west, the nation's flocks expanded, but a shortage of sheep bearing low-grade wool to clothe enslaved people led to huge imports of cheap wool, and the imposition of tariffs on this contributed to the rift between the North and the South. Sheep were an integral part of westward expansion, even though popular perception sees the old West as a land of cattle. We saw how in the West, an important step toward conservation was setting lands aside where sheep were not permitted. In an industrializing America, slaughtered sheep supplied Europeans with mutton while by-products of the slaughter became soap and violin strings. Sheep have been involved in everything from taxation of income to lawn mowing and dog licensing. In the development of the United States, sheep have mattered enormously.

EPILOGUE

"The Sheep Follow Him, for They Know His Voice"

> So the Lord blessed the latter end of Job more than his beginning:
> for he had fourteen thousand sheep.—Job 42:12

Thomas Jefferson found the ethical teachings of Jesus very appealing, but the New Testament contained other passages he viewed as far less insightful or enlightening. So Jefferson took a Bible, cut out only the sections that he admired, and pasted them into a blank book in an order he thought appropriate. This edited compilation is commonly called "the Jefferson Bible."[1]

Jefferson's scriptural collage includes several references to sheep—not surprising, given the prominence of ovines in the ministry of Jesus. Thus, chapter 32 of the Jefferson Bible is a portion of the Parable of the Good Shepherd, John 10:1–18. In part it reads:

> Verily, verily, I say unto you, He that entereth not by the door into the sheepfold, but climbeth up some other way, the same is a thief and a robber. But he that entereth in by the door is the shepherd of the sheep. To him the porter openeth; and the sheep hear his voice: and he calleth his own sheep by name, and leadeth them out. And when he putteth forth his own sheep, he goeth before them, and the sheep follow him: for they know his voice. And a stranger will they not follow, but will flee from him: for they know not the voice of strangers.

When the workday at the Atlanta History Center is near its end and it is time for me to bring the sheep into the barn for the evening, I call them by their names. They come inside the barn in a particular order, the ones that occupy the stall farthest from the door being the first in line to scoot across the threshold. Often visitors viewing this procedure express astonishment.

"They know their names? You can call them by their names and they will respond?" Yes, they will. It is just as John the Apostle described it so many years ago.

I can't help but imagine that as Jefferson read the Parable of the Good Shepherd and decided it was appropriate for his compilation, he thought of the significance of sheep in his own life and of their great importance to all the citizens of the country he helped to found.

While in my daily duties I use my voice to summon sheep by their names, in *American Sheep: A Cultural History* I have tried to use the written word to bring to life the people and events that made sheep such a significant part of the nation's history. I hope I have shown that when we look closely at the growth and development of the United States, we are never very far from the story of America's shepherds, calling their sheep. Those sheep followed because they knew those familiar voices.

List of Abbreviations

Annals Cong.	Annals of Congress
Ann. Rep. Central Park	*Annual Report of the Board of Commissioners of the Central Park* (New York: Bryant)
Cong. Globe	Congressional Globe
Cong. Rec.	Congressional Record
JCC	*Journals of the Continental Congress 1774–1789* (Washington, D.C.: Government Printing Office, 1904)
NAF	Founders Online, National Archives, https://founders.archives.gov
PAF	Lyman H. Butterfield, et al., eds., *The Adams Papers, Adams Family Correspondence* (Cambridge, Mass.: Harvard University Press, 1963–)
PAH	Harold C. Syrett and Jacob E. Cooke, eds., *The Papers of Alexander Hamilton* (New York: Columbia University Press, 1961–1987)
PBF	Leonard W. Labaree, et al., eds., *The Papers of Benjamin Franklin* (New Haven, Conn.: Yale University Press, 1959–)
PGW	Donald Jackson, et al., eds., *The Papers of George Washington* (Charlottesville: University of Virginia Press, 1976–1999)
PTJ	Julian P. Boyd, et al., eds., *The Papers of Thomas Jefferson* (Princeton, N.J.: Princeton University Press, 1950–)
Reg. Debates	Register of Debates (U.S. Congress)
Rep. Finances	Report of the Secretary of the U.S. Treasury on Finances (Washington, D.C.: Government Printing Office) dates indicated
U.S. Stats.	United States Statutes at Large

Notes

Preface. Dana's Invisible Ancestors

1. Here and throughout, figures on the number of sheep in America are from the U.S. Census unless otherwise indicated.

Introduction

1. George Dangerfield, *Chancellor Robert R. Livingston of New York, 1746–1813* (New York: Harcourt, Brace, 1960).

2. Much of the *Essay* was written in 1806, but Livingston rewrote and enlarged it three years later; Dangerfield, *Chancellor Robert R. Livingston,* 435. Dangerfield lauds *Essay on Sheep* as "a remarkable performance" ("it offers, in the flights of its speculation no less than in its marshaling of fact, an admirable example of the thinking of the American Enlightenment"), 436.

3. "To James Madison from Robert R. Livingston, 17 January 1809," Founders Online, National Archives, hereafter NAF, https://founders.archives.gov/documents/Madison/99-01-02-1305.

4. Anonymous review of *Essay on Sheep, Medical Repository,* Third Hexade, no. 1 (1810), 268. The review likely was written by Samuel Latham Mitchill, founder and editor of the periodical. See Alan David Aberbach, *In Search of an American Identity: Samuel Latham Mitchill, Jeffersonian Nationalist* (New York: Peter Lang, 1988), ix.

5. Eric Jay Dolin, *Fur, Fortune, and Empire: The Epic History of the Fur Trade in America* (New York: Norton, 2010), 103.

6. Ross King, *Brunelleschi's Dome: How a Renaissance Genius Invented Architecture* (New York: Penguin, 2000), 2; Alan Butler, *Sheep* (Ropley, UK: O Books, 2006), 65–73.

7. Butler, *Sheep,* 84, notes that the money to fund Columbus's journey came from Spain's wool profits.

8. PGW, Presidential Series, 20:626–630.

9. Charles Hudson, *The Southeastern Indians* (Knoxville: University of Tennessee Press, 1976), 262–263.

10. Francis Guy, *Winter Scene in Brooklyn*, 1820, oil on canvas (Crystal Bridges Museum of American Art 2006.98). A similar Guy painting with the same title is held at the Brooklyn Museum of Art.

11. On the importance of horses in pre-automobile America, see Ann Norton Greene, *Horses at Work: Harnessing Power in Industrial America* (Cambridge, Mass.: Harvard University Press, 2008); Clay McShane and Joel Tarr, *The Horse in the City: Living Machines in the Nineteenth Century* (Baltimore: Johns Hopkins University Press, 2007).

12. See chapter 8.

13. Robert R. Livingston, *Essay on Sheep* (New York: T. & J. Swords, 1809), 20–21. In making this point, Livingston was adopting the four-stage theory of civilization, popular in his time. The theory viewed mankind as progressing in phases from hunter-gatherers to pastoral nomads, then farming agriculturists, and finally to commercial manufacturers. It was first proposed by the eighteenth-century Scotch legal theorist Lord Kames. See Arthur Herman, *How the Scots Invented the Modern World* (New York: Three Rivers, 2001), 98–103. The four-stage theory eventually fell out of favor, but more recently scientists have argued that domestication was a two-way process; as the animals involved were altered, humans themselves were correspondingly changed. See Raymond Pierotti and Brandy R. Fogg, *The First Domestication: How Wolves and Humans Coevolved* (New Haven: Yale University Press, 2017); and Richard C. Francis, *Domesticated: Evolution in a Man-Made World* (New York: Norton, 2015).

Chapter 1. There Once Were Some Sheep from Nantucket

1. Andrew Moore, "The American Farmer as French Diplomat: J. Hector St. John de Crèvecoeur in New York after 1783," *Journal of the Western Society for French History* 39 (2011): 133–143.

2. Ibid., 134.

3. Ibid., 133.

4. J. Hector St. John de Crèvecoeur, *Letters from an American Farmer* (1782; repr., New York: Duffield, 1904), 124. There is discussion among scholars about whether *Letters from an American Farmer* is a straightforward account or a form of epistolary novel or romance. See Nathaniel Philbrick, "The Nantucket Series in Crèvecoeur's 'Letters of an American Farmer,'" *New England Quarterly* 64, no. 3 (September 1991): 414; Robert P. Winston, "'Strange Order of Things!': The Journey

to Chaos in 'Letters from an American Farmer,'" *Early American Literature* 19, no. 3 (Winter 1984–85): 249. The possibility that this is a fictionalized work would make it a questionable source for information on the sheep of Nantucket and Martha's Vineyard were there not other documentation of the flocks on these islands.

5. Crèvecoeur, *Letters from an American Farmer*, 124.

6. Ibid., 130–132.

7. Ibid., 137, 166.

8. Alfred W. Crosby Jr., *Ecological Imperialism: The Biological Expansion of Europe, 900–1900* (Cambridge: Cambridge University Press, 1986); and Crosby, *The Columbian Exchange: Biological and Cultural Consequences of 1492* (Westport, Conn.: Praeger, 2003).

9. Crosby, *Ecological Imperialism*, 190–192.

10. Ibid., 71–73.

11. Gulf Coast or Gulf Coast Native Sheep," Livestock Conservancy, accessed October 9, 2023, https://livestockconservancy.org/heritage-breeds/heritage-breeds-list/gulf-coast-sheep; "History," Gulf Coast Sheep Breeders Association, accessed October 9, 2023, http://www.gulfcoastsheep.info/history.

12. William Cronon, *Changes in the Land: Indians, Colonists, and the Ecology of New England* (New York: Hill & Wang, 1983), 129; Edmund S. Morgan, *American Slavery, American Freedom: The Ordeal of Colonial Virginia* (New York: Norton, 1975), 87; Virginia DeJohn Anderson, *Creatures of Empire: How Domestic Animals Transformed Early America* (Oxford: Oxford University Press, 2004), 110, 147.

13. Anderson, *Creatures of Empire*, 110.

14. Ibid., 110, 147; Chester Whitney Wright, *Wool Growing and the Tariff: A Study in the Economic History of the United States* (Boston: Houghton Mifflin, 1910), 2. In contrast to the colonial situation, wolves had been eradicated from England by the fifteenth century. See Keith Thomas, *Man and the Natural World: Changing Attitudes in England 1500–1800* (New York: Oxford University Press, 1983), 273. Of the elimination of English wolves, Thomas writes: "It made English sheep-farming less labor-intensive, for shepherds no longer had to guard their flocks by night ... or lock them up in stone sheepcotes." The early English colonists in North America would enjoy no such advantages in their shepherding.

15. Anderson, *Creatures of Empire*, 147–148.

16. Ibid., 148; Wright, *Wool Growing and the Tariff*, 2.

17. For example, Pauline Maier, *From Resistance to Revolution: Colonial Radicals and the Development of American Opposition to Britain, 1765–1776* (New York: Norton, 1991), 51–76; and John Ferling, *A Leap in the Dark: The Struggle to Create the American Republic* (New York: Oxford University Press, 2003), 23–52.

18. Ferling, *A Leap in the Dark*, 31–32.

19. PBF, 13:133.

20. Ibid., 137.

21. Ibid., 140.

22. Ibid., 140–141.

23. PBF 11:357–359.

24. PBF 12:132–135.

25. *Journals of the Continental Congress 1774–1789* (Washington, D.C.: Government Printing Office, 1904), hereafter JCC, 1:13–14.

26. Ibid., 78. The colonists' drive to divest from British wool imports and to improve the American sheep began at the local level several years earlier. A 1768 almanac in New England urged Massachusetts residents, "[Do not] purchase any sort of Woolen Goods made abroad for 12 to 18 Months to come, but ... wear ... old patch'd Cloaths, till our own Manufacture can be brought, as many in New York, Connecticut, and Philadelphia are now doing." Another New England almanac in 1870 urged citizens, "Go on as you have wisely begun, to increase your flocks of Sheep." See Chester Noyes Greenough, "New England Almanacs, 1766–1775 and the American Revolution," *Proceedings of the American Antiquarian Society* 45, no. 2 (October 1935), 298, 304.

27. Gordon S. Wood, *The Creation of the American Republic: 1776–1787* (Chapel Hill: University of North Carolina Press, 1998) 68.

28. Bruce E. Steiner, *Samuel Seabury, 1729–1796: A Study in the High Church Tradition* (Athens: Ohio University Press, 1971); Ann W. Rowthorn, *Samuel Seabury: A Bicentennial Biography* (New York: Seabury, 1983).

29. Samuel Seabury, *A View of the Controversy between Great-Britain and Her Colonies: Including a Mode of Determining Their Present Disputes, Finally and Effectually; and of Preventing All Future Contentions.* (New York, 1775), 57–58.

30. PAH, 1:45–78.

31. Andrea Wulf, *Founding Gardeners: The Revolutionary Generation, Nature, and the Shaping of the American Nation* (New York: Knopf, 2011), 113–119.

32. PAF 10:109–110.

33. PAF 10: 94–96.

34. Ibid, 97.

35. Ibid.

36. PGW, Revolutionary War Series, 1:262–264 (British taking two thousand sheep from Fisher's Island, N.Y.); ibid., 1:447–448 (British forces "suffer greatly for want of fresh Provisions notwithstanding they have pillaged several Islands of a good many Sheep and Cattle"); and chap. 1, note 1 above.

37. PGW 13:111.

38. See, for example, Ferling, *A Leap in the Dark*, 217.

39. NAF, https://founders.archives.gov/documents/Washington/99-01-02-12012.

See also John Ferling, *Almost a Miracle: The American Victory in the War of Independence* (New York: Oxford University Press, 2007).

40. Robert R. Livingston, *Essay on Sheep* (New York: T. & J. Swords, 1809), 53; Wright, *Wool Growing and the Tariff*, 35–37.

41. Rebecca J. H. Woods, *The Herds Shot Round the World: Native Breeds and the British Empire, 1800–1900* (Chapel Hill: University of North Carolina Press, 2017), 54.

42. John Sinclair, *Address to the Society for the Improvement of British Wool* (Edinburgh: Society for the Improvement of British Wool, 1791), 6–7. See also John Lord Somerville, *Facts and Observations Relative to Sheep, Wool, Ploughs, and Oxen* (London: John Harding, 1809), 1–2, advocating sheep improvement: "The political situation of Spain may be such to shut out, or at least materially increase, the present difficulty of importing her wools into this country, in which case, it is a matter of the utmost national importance, that the fine-woolen trade of Great Britain should suffer nothing in reputation."

43. For more on this, see Woods, *The Herds Shot Round the World*.

Chapter 2. Embargos, Merinos, and the Flocks Grow

1. *Annals Cong.* 15 (1806), 450.
2. Gordon S. Wood, *Empire of Liberty: A History of the Early Republic, 1789–1815* (New York: Oxford University Press, 2009) 639–646.
3. *Annals Cong.* 15 (1806), 450–451.
4. Ibid.
5. *U.S. Stats.* 2(1806), 379.
6. Charles Darwin, *The Origin of Species* (New York: Modern Library, 1998), 50–51.
7. Ibid., 53.
8. Genesis 30:31–43.
9. Phillip Armstrong, *Sheep* (London: Reaction, 2016), 60–61. See also J. D. Pearson, "A Mendelian Interpretation of Jacob's Sheep," *Science and Christian Belief* 13, no. 1 (2001): 51–58. The biblical account declares that Jacob has set up branches with a streaked appearance and that this visual stimulus causes the ewes to bear spotted lambs—a scientific impossibility. Pearson makes a solid case that the branches mentioned in the story were actually used to make a fence so to keep undesirable rams from mating with ewes, a basic requirement of a selective breeding program.
10. Richard Francis, *Domestication: Evolution in a Manmade World* (New York: Norton, 2015), 166.
11. PGW, Presidential Series, 13:9. If Washington's sheep really produced that

much wool, this is notable, since the two and a half pounds per sheep Whitting reported was not an especially meager yield for the time.

12. PGW, Presidential Series, 12:523–529. Washington's glowing assessment of his own sheep was not shared by Richard Parkinson, a British agricultural writer who toured the United States and called on Washington at Mount Vernon. Parkinson wrote: "The General's opinion of his own land, cattle, sheep, &c., was not at all like that of a man of information. His sheep were very shabby ones: the wool from his sheep at the time of clipping, would not average more than three pounds a fleece. He told me his sheep were much better before the war, and pleaded want of care." Richard Parkinson, *The Experienced Farmer's Tour in America* (London, 1805), 426–427. See also PGW, Retirement Series, 1:323–325.

13. PGW, Presidential Series, 12:610–616.

14. Annals Cong. 1 (1790), 969.

15. Annals Cong. 1 (1790), 1095.

16. John Ferling, *Jefferson and Hamilton: The Rivalry That Forged a Nation* (New York: Bloomsbury, 2013) 67–68.

17. PAH 10:291.

18. See editorial notes in PAH 10:230–340. The editorial notes to "Report on Manufacture'" (PAH 10:230–340) give several additional examples of Smith's influence on Hamilton. Other Founders were similarly influenced by *The Wealth of Nations*. "Most public men in America acquired at least a passing acquaintance with the work, almost all praised it, and many gave it thorough study." Forrest McDonald, *Novus Ordo Seclorum: The Intellectual Origins of the Constitution* (Lawrence: University Press of Kansas, 1985), 128.

19. Adam Smith, *The Wealth of Nations*, edited by Edwin Cannan (New York: Bantam Classic, 2003), 312.

20. Ibid., 822–823.

21. Ibid., 313.

22. Ibid., 827.

23. Ibid., 830.

24. PAH 10:332.

25. Ibid. Earlier in the report Hamilton distinguished *premiums* from *bounties*, which he described as government support "applicable to the whole quantity of an article produced, or manufactured, or exported" rather than limited to excellence in the field, as premiums were. Hamilton suggested bounties as an appropriate means to increase foreign interest in supplying American wool needs.

26. Ron Chernow, *Alexander Hamilton* (New York: Penguin, 2004), 378.

27. U.S. Constitution, article 1, section 8; PAH 10:302–303.

28. NAF, https://founders.archives.gov/documents/Jefferson/99-01-02-2842.

29. NAF, https://founders.archives.gov/documents/Jefferson/99-01-02-2859.

30. U.S. Constitution, article 1, section 8.

31. Ibid., article 2, section 2.

32. Ferling, *Jefferson and Hamilton*, 218–221.

33. NAF, http://founders.archives.gov/documents/Jefferson/99-01-02-2859.

34. Carroll W. Pursell Jr., "E. I. du Pont, Don Pedro, and the Introduction of Merino Sheep into the United States, 1801: A Document," *Agricultural History* 33, no. 2 (April 1959): 86–88; Carroll W. Pursell Jr., "E. I. du Pont and the Merino Mania in Delaware, 1805–1815," *Agricultural History* 36, no. 2 (April 1962): 91–100.

35. Julius Klein, *The Mesta: A Study in Spanish Economic History 1273–1836* (Cambridge: Harvard University Press, 1920), 3–16; Livingston, *Essay on Sheep*, 35–36; Rebecca J. H. Woods, "Green Mountain Merinos: From New England to New South Wales in the Nineteenth Century," *Vermont History* 85, no. 1 (Winter/Spring 2017): 1–19, 4. While Spain tried to prevent exports of live sheep, the wool from their Merinos was shipped to England for manufacture into cloth, as noted earlier in the chapter.

36. Pursell, "E.I. du Pont, Don Pedro, and the Introduction of Merino Sheep," 87; Woods, "Green Mountain Merinos," 4–5; Ezra A. Carman, H. A. Heath, and John Minto, *Special Report on the History and Present Condition of the Sheep Industry of the United States* (Washington, D.C.: Government Printing Office, 1892), 132.

37. PTJ, Original Series, 38:200–201, editorial note.

38. Woods, "Green Mountain Merinos," 8.

39. Ibid., 11. Livingston insisted that crossing a Merino ram with a ewe of any type would increase the value of the resulting offspring's wool by at least a third. See *Essay on Sheep*, 166.

40. Pursell, "E. I. du Pont and the Merino Mania in Delaware," 96.

41. Richard Peters, "Tunis, Broad-tailed Mountain Sheep," *Memoirs of the Philadelphia Society for Promoting Agriculture* 2 (1811): 211, paper read before the organization May 8, 1810. See chap. 5 for more on Judge Peters.

42. Livingston, *Essay on Sheep*, 63–62. Custis was George Washington's adopted son. Livingston himself was lauded for his sheep-related patriotism in the review of his book that appeared in the *Medical Repository*, Third Hexade, no. 1 (1810), 268.

43. PTJ, Retirement Series, 2:165–166.

44. Edward Gibbon, *The Decline and Fall of the Roman Empire*, abridged ed. (New York: Modern Library, 2005), 16. Thomas Jefferson viewed Gibbon's work favorably; John Adams did not. See PTJ, Retirement Series, 1:580–582; and NAF, https://founders.archives.gov/documents/Adams/99-02-02-6799.

45. See Andrew Burstein and Nancy Isenberg, *Madison and Jefferson* (New York: Random House, 2010).

46. PTJ, Retirement Series 2:388–390. See also Jefferson's 1810 letter to the architect William Thornton, designer of the U.S. Capitol: "On the subject of the Merinos . . . I confess to you that I have not been satisfied with that kind of patriotism the strongest feature of which is to enrich the patriot himself." PTJ, Retirement Series 2:412–414.

47. JCC, 1:78.

48. PTJ, Retirement Series 2: 388–390. Madison responded favorably to Jefferson's idea, telling his mentor, "The general idea of disposing of the supernumerary Merino Rams for the public benefit had occurred to me. The mode you propose for the purpose seems well calculated for it." PTJ, Retirement Series 2:418–419.

49. See, for example, Gordon S. Wood, *Empire of Liberty: A History of the Early Republic, 1789–1815* (New York: Oxford University Press, 2009), 660 ("From beginning to end the war seemed . . . ludicrous"); and Sean Wilentz, *The Rise of American Democracy: Jefferson to Lincoln* (New York: Norton, 2005), 142 ("The war exposed the nation's internal rifts and military weaknesses").

50. PTJ, *Retirement Series,* 7:13–16. See also PTJ, *Retirement Series,* 4:637–638.

51. Elkanah Watson, *History of Agricultural Societies on the Modern Berkshire System* (Albany, N.Y.: Steele, 1820), 116.

52. Ibid.

53. Ibid, 120.

54. Mark A. Mastromarino, "Fair Visions: Elkanah Watson (1758–1842) and the Modern American Agricultural Fair" (PhD diss., College of William and Mary, 2002), 182.

55. Ibid., 90, 157. Mastromarino also writes that Watson's works have been treated by some historians as unbiased primary sources; for an example of this, see Ulysses Prentiss Hedrick, *A History of Horticulture in America to 1860* (New York: Oxford University Press, 1950), 511–512.

56. Mastromarino, "Fair Visions," 219–223.

57. Ibid., 215.

58. Ibid., 72, 144, 156, 309.

59. For sources on Taylor and his influence, see Garrett Ward Sheldon and C. William Hill Jr., *The Liberal Republicanism of John Taylor of Caroline* (Madison, N.J.: Farleigh Dickinson University Press, 2008); Robert E. Shalhope, *John Taylor of Caroline: Pastoral Republican* (Columbia: University of South Carolina Press, 1980); and M. E. Bradford, "A Virginia Cato: John Taylor of Caroline and the Agrarian Republic," introduction to John Taylor, *Arator: Being a Series of Agricultural Essays, Practical and Political; in Sixty-four Numbers* (1818; repr., Indianapolis: Liberty Classics, 1977), 11–43. Sheldon and Hill assert, "Recent scholarship in early American political thought has raised John Taylor of Caroline to new prominence" (11).

60. Bradford, "A Virginia Cato," 13.

61. Shalhope, *John Taylor of Caroline*, 127. Shalhope notes that earlier historians inaccurately thought that the essays first appeared in 1803 (220–221).

62. Sheldon and Hill, *The Liberal Republicanism of John Taylor of Caroline*, 69.

63. Lynn A. Nelson, *Pharsalia: An Environmental Biography of a Southern Plantation, 1780–1880* (Athens: University of Georgia Press, 2007), 75.

64. Shalhope, *John Taylor of Caroline*, 3.

65. Nelson, *Pharsalia*, 79.

66. Taylor, *Arator*, 179.

67. Ibid., 179–80.

68. Ibid., 181. Taylor's insistence that the United States should rely on foreign rather than homegrown sources for wool is reflected in his opposition to protective tariffs. See the following chapter in this book; and Taylor, *Arator*, 74–76. See also Taylor's letter in *Memoirs of the Philadelphia Society for Promoting Agriculture* 1 (1808): 331.

69. PTJ 28, Original Series, 293–301. Jefferson and Taylor corresponded regularly, sometimes on politics but more frequently on farming practices. See Shalhope, *John Taylor of Caroline*, 100–101.

70. Taylor's other points were more dubious, that cattle were a superior source of manure and that they provided dairy. Others of Taylor's time asserted that sheep were unmatched for manuring, and some sheep are used for milking, even if this has never been a particularly common use for them in the United States. On sheep for manuring, see Henry S. Randall, *Sheep Husbandry in the South* (Philadelphia: J. S. Skinner, 1848), 72–73. On sheep for dairy, see Richard C. Francis, *Domestication: Evolution in a Man-Made World* (New York: Norton, 2015), 164–65.

71. Taylor, *Arator*, 225–226.

72. See, for example, John S. Skinner, ed., *The Monthly Journal of Agriculture*, vol. 1 (New York: Greeley & McElrath, 1846), 38–39; Randall, *Sheep Husbandry in the South*, 72; Thomas Pollard, *Sheep Husbandry for Virginia* (Richmond, Va.: R. F. Walker, Superintendent of Public Printing, 1881), 3–6.

73. See, for example, Nelson, *Pharsalia*, 159, 188, 217.

74. See Sven Beckert, *Empire of Cotton: A Global History* (New York: Knopf, 2015).

Chapter 3. Utter Destruction to the Prosperity

1. George P. Rawick, ed., *The American Slave: A Composite Autobiography*, supplement, series 1, vol. 3 (Westport, Conn.: Greenwood, 1977), 177, 181.

2. Helen Bradley Foster, *New Raiments of Self: African American Clothing in the Antebellum South* (Oxford, U.K.: Berg, 1997), 85–86.

3. Rawick, *The American Slave*, supplement, series 1, vol. 1, 301.

4. See *Born in Slavery: Slave Narratives from the Federal Writers' Project, 1936 to 1938*, Library of Congress, https://www.loc.gov/collections/slave-narratives-from-the-federal-writers-project-1936-to-1938/about-this-collection. For brief accounts of all the narratives in which the former enslaved mentioned wearing wool, see Foster, *New Raiments of Self*, 85–86.

5. PTJ, Retirement Series, 1:666–667.

6. PTJ, Retirement Series, 4:428–430. Jefferson did admit, "For fine stuff we shall depend on your Northern manufactures." Those manufacturers would soon not only be producing "fine stuff" but also woolens at the opposite end of the quality spectrum.

7. Thomas Jefferson, *Notes on the State of Virginia*, edited by William Peden (Chapel Hill: University of North Carolina Press, 1954), 165.

8. Steven Hahn, *The Roots of Southern Populism: Yeoman Farmers and the Transformation of the Georgia Upcountry 1850–1890* (New York: Oxford University Press, 1983), 30.

9. "Wright, Silas, Jr.," Biographical Directory of the United States Congress, http://bioguide.congress.gov/scripts/biodisplay.pl?index=W000770.

10. Reg. Debates 20th Cong., 1st sess.(1828), 1838. Wright inadvertently called his derived figure a count of sheep in the whole United States, but it is clear he only meant it as an estimate for the northeastern states.

11. Ibid. Another New York representative, Michael Hoffman, also speculated on the nation's sheep population. See Reg. Debates, 20th Cong., 1st sess., 1978–79 (1828).

12. Chester Whitney Wright, *Wool-Growing and the Tariff: A Study in the Economic History of the United States* (Cambridge, Mass.: Harvard University Press, 1910), 71.

13. Tench Coxe, *Tabular Statement of the Several Branches of American Manufactures* (Philadelphia: Cornman, 1813), table 46. The states reporting their sheep were Massachusetts, Vermont, Connecticut, New Jersey, and Pennsylvania, plus the Michigan Territory. Pennsylvania reported the most sheep, 618,223.

14. Tench Coxe, *A Statement of the Arts and Manufactures of the United States of America* (Philadelphia: Cornman, 1814), xiv.

15. C. Benton and S. F. Barry, *A Statistical View of the Number of Sheep* (Cambridge, Mass.: Folsom, Wells, & Thurston, 1837).

16. Ibid., v.

17. Ibid., 53–54.

18. Ibid., 56. Apparently only wool figured in this calculation; there was no mention of mutton.

19. Ibid., 106. Their exact figure was 12,897,638.

20. PAH, 9:332–334.

21. U.S. Census 1790, 3.

22. Adam Beatty, "Sheep Husbandry in Kentucky," *American Agriculturist* 4, no. 5 (May 1845) 146–147. Beatty's statistics were cited favorably by sheep husbandry books of the era. See L. A. Morrell, *The American Shepherd: Being a History of the Sheep* (New York: Harper, 1863) 152; Randall, *Sheep Husbandry in the South*, 127.

23. Randall, *Sheep Husbandry in the South*, 127. While Randall's remark that enslaved Americans consumed less wool than free citizens is unassailable, the figures he presents are confusing. Inexplicably, he writes that a northern farmer or laborer "in comfortable circumstances" would consume about twenty pounds of wool per year, whereas an enslaved person would receive only eight to ten pounds. In fact, such a consumption of wool would far exceed the available supply, both domestically and from imports.

24. Benton and Barry, *A Statistical View of the Number of Sheep*, iii.

25. Robert V. Remini, *Henry Clay: Statesman for the Union* (New York: Norton, 1991), 330.

26. U.S. Stats. 4 (1828), 270–275.

27. Daniel Walker Howe, *What God Hath Wrought: The Transformation of America, 1815–1848* (New York: Oxford University Press, 2007), 274; Wilentz, *The Rise of American Democracy*, 299; Robert V. Remini, "Martin van Buren and the Tariff of Abominations," *American Historical Review* 63, no. 4 (July 1958): 906. The political maneuverings leading to the 1828 tariff, as well as its unintended consequences, are beyond the scope of this book. I will simply note that economic historian William Taussig commented that the 1828 tariff was passed "in a form approved by no one." Frank William Taussig, *The Tariff History of the United States*, 5th ed. (New York: Putnam, 1910), 90.

28. In the debate, Congressman John Barney of Maryland enumerated the few railroads in operation at the time. Reg. Debates, 4:2247 (1828).

29. U.S. Stats. 4 (1828), 271.

30. Reg. Debates 4 (1828), 2354 (comments of Edward Bates). See also ibid., 2054–55 ("We have spent nearly four weeks in debating the subject of wool and woolens"; comments of Peleg Sprague). In contrast, there wasn't as much talk in the Senate about sheep and their wool. One of the few to address the subject was Daniel Webster of Massachusetts who opined, "The interest of the sheep owner is the great object which the bill is calculated to benefit." Reg. Debates 4 (1828), 758.

31. For details on the 1827 maneuverings, see Edward Stanwood, *American Tariff Controversies in the Nineteenth Century*, vol. 1 (Boston: Houghton Mifflin, 1903), 255–261; Andrew W. Young, *A History of the American Protective System and Its Effects upon the Several Branches of Domestic Industry* (New York: Miller, 1864), 170–196.

32. Reg. Debates 3 (1827), 864–865.

33. Ibid., 943.

34. Ibid., 964. See also the remark of Charles Miner of Pennsylvania that no part of the world was better for raising sheep than the northern part of his state, 1076.

35. Ibid., 781.

36. Reg. Debates 4 (1828),2276.

37. Ibid., 2269. See also Wright, *Wool-Growing and the Tariff*, 65. Wright calls the three classes "coarse, common, and superfine."

38. Reg. Debates 4 (1828),2269. The city's name was spelled "Buenos Ayres" in these published debates; here and throughout I have modernized the spelling. Other representatives agreed that the United States didn't produce any of this coarse wool. See remarks of Isaac Bates of Massachusetts (2008); Andrew Stewart of Pennsylvania (2224); and Andrew Wright of Ohio (2286). See also Wright, *Wool-Growing and the Tariff*, 103.

39. American State Papers, Finance, 5, (1828),805.

40. Ibid., 805.

41. Ibid., 806, 828. In Congress, several representatives also spoke of factories producing Negro cloths, see Reg. Debates, 4:1936, 2007, 2076 (1828), remarks of Daniel Barnard, Isaac Bates, and Peleg Sprague, respectively.

42. Taussig, *The Tariff History of the United States*, 71, 81–82.

43. *Niles Weekly Register*, 24 (March 22, 1823), 40. See also American State Papers, Commerce and Navigation, 2 (1822), 612; and American State Papers, Finance, 5 (1827), 602, 604.

44. Randall, *Sheep Husbandry in the South*, 124. Randall cites "Reports of the Secretary of the Treasury" as the primary source for his data. I have examined the Treasury Department's annual reports of the 1840s and have been unable to discover the document he is referring to, which apparently gave a breakdown of the countries of origin of the imported wool. Perhaps through his connections Randall was able to examine unpublished records. The notion that Randall was presenting authentic figures is buttressed, however, since his figures for total wool imports match those in documents that I have examined: Rep. Finances, 34th Cong., 3rd sess., S. Doc. No. 3 (1856) at 203; and *Reports of the Secretary of the Treasury of the United States*, vol. 5. (Washington, D.C.: John C. Rives, 1851) 30. See also Wright, *Wool-Growing and the Tariff*, 99, 340. His figures are similar to Randall's although not exact.

45. "The Statue of Liberty," Ellis Island Foundation, Inc.; https://www.libertyellisfoundation.org/statue-facts.

46. Cong. Globe, 39th Cong., 1st sess. (1866), appendix, 292.

47. Randall, *Sheep Husbandry in the South*, 124.

48. Michael Allaby, ed., *The Concise Oxford Dictionary of Zoology* (Oxford: Oxford University Press, 1991).

49. Reg. Debates 4 (1828), 2008. See also 2287, remarks of John Wright of Ohio.

50. Randall, *Sheep Husbandry in the South*, 52.

51. Ibid., 157.

52. Reg. Debates 4 (1828), 1867. In spite of Wright's declaration, given the huge amount of cheap wool imported by the 1840s no doubt some of it wound up as clothing or blankets for poor white Americans. But unquestionably the primary reason for these imports was to facilitate the desire of enslavers to clothe their laborers at the lowest cost possible.

53. Ibid., 2444.

54. Reg. Debates 4 (1828), 2444. Similar comments were made in the Senate by Nathaniel Macon of North Carolina. See Reg. Debates, 4:769.

55. Reg. Debates 8 (1832), 1620.

56. Wright, *Wool-Growing and the Tariff*, 61; Taussig, *The Tariff History of the United States*, 41.

57. U.S. Stats. 3 (1816), 310–314; Wright, *Wool-Growing and the Tariff*, 41. Wright presents the figure incorrectly in the chart in the appendix, 344.

58. U.S. Stats. 3 (1816), 313. Notably, wool was not subject to a tariff if it was still attached to a living sheep; among the things the 1816 law left duty-free were "animals imported for breed." Thus, those working to augment America's sheep with Merinos and other foreign varieties had no tax burden.

59. U.S. Stats. 3 (1816), 310.

60. *Niles Weekly Register* 24 (March 22, 1823), 40; American State Papers, Commerce and Navigation, 2 (1822),612 (1822); Reg. Debates 4 (1828),2360 (comments of Churchill Cambreleng).

61. American State Papers, Commerce and Navigation, 2 (1822), 612 (1822).

62. Wilentz, *The Rise of American Democracy*, 206.

63. Wright, *Wool-Growing and the Tariff*, 41.

64. Taussig, *The Tariff History of the United States*, 28.

65. U.S. Stats. 4 (1824), 25–26.

66. Ibid., 26.

67. U.S. Stats. 4 (1828), 271.

68. Reg. Debates 4 (1828),1768.

69. Reg. Debates 8:3293 (1832).

70. Ibid., 3511. See similar remarks by Ingersoll of Connecticut at 3665.

71. William Lee Miller, *Arguing about Slavery: John Quincy Adams and the Great Battle in the United States Congress* (New York: Vintage, 1995); Marcus Rediker, *The Amistad Rebellion: An Atlantic Odyssey of Slavery and Freedom* (New York: Viking, 2012).

72. Wright, *Wool-Growing and the Tariff*, 51.

73. U.S. Stats. 4 (1832),583.

74. Wright, *Wool-Growing and the Tariff*, 84.

75. "South Carolina Ordinance of Nullification, November 24, 1832," Avalon Project, Yale Law School, https://avalon.law.yale.edu/19th_century/ordnull.asp. On the causes of and reaction to this action, see William W. Freehling, *Prelude to Civil War: The Nullification Controversy in South Carolina 1816–1836* (New York: Oxford University Press, 1965); Donald J. Ratcliffe, "The Nullification Crisis, Southern Discontents, and the American Political Process," *American Nineteenth Century History* 1, no. 2 (April 2000): 1–30; and David P. Currie, *The Constitution in Congress: Democrats and Whigs, 1829–1861* (Chicago: University of Chicago Press, 2005), 99–117.

76. "South Carolina Ordinance of Nullification, November 24, 1832." South Carolina planters did not directly pay the tariff on raw wool. The duty was paid by the importers, but the increased cost was passed along to the manufacturers and later to the slaveholding consumers.

77. See Edward E. Baptist, *The Half Has Never Been Told: Slavery and the Making of American Capitalism* (New York: Basic, 2014); and Beckert, *Empire of Cotton*.

78. Studies of antebellum plantations with sheep typically report either that the animals produced fine wool or that they were kept for mutton. See Lynn A. Nelson, *Pharsalia: An Environmental Biography of a Southern Plantation 1780–1880* (Athens: University of Georgia Press, 2007), 159; Drew A. Swanson, *Remaking Wormsloe Plantation: The Environmental History of a Lowcountry Landscape* (Athens: University of Georgia Press, 2012), 81; and Sam Bowers Hilliard, *Hog Meat and Hoecake: Food Supply in the Old South 1840–1860* (Athens: University of Georgia Press, 2014), 45–46.

79. American State Papers, Finance, 2:427 (1810).

80. Wright, *Wool-Growing and the Tariff*, 19.

81. Ibid., 17.

82. PTJ, Retirement Series, 5:129–130.

83. Ibid., 5:130–131.

84. Ibid., 5:448–449.

85. Jefferson, *Notes on the State of Virginia*, 165.

86. Ibid., 165.

87. David P. Currie, *The Constitution in Congress: The Jeffersonians, 1801–1829* (Chicago: University of Chicago Press, 2001) 283–289.

88. Howe, *What God Hath Wrought*, 397.

89. Reg. Debates, 4:2445 (1828).

90. *Journal of the Senate of South Carolina 1861* (Columbia: Pelham, 1861), 45.

91. Ibid., 154; *Journal of the House of Representatives of the State of South Carolina 1861* (Columbia: Pelham, 1861), 294.

92. U.S. Census 1860, Agriculture, cviii–cix.

93. Wright, *Wool-Growing and the Tariff*, 172–175.

94. Ibid., 176.

95. *Soldiers' National Cemetery at Gettysburg, Pennsylvania, Indiana* General Assembly, doc. no. 17, pt. 2 (Indianapolis: Holloway, 1865), 600.

96. Gordon L. Jones, *Confederate Odyssey: The George W. Wray Jr. Civil War Collection at the Atlanta History Center* (Athens: University of Georgia Press, 2014), 358.

Chapter 4. Shepherds Enslaved and Free

1. A. W. Wayman, *My Recollections of African M.E. Ministers, or Forty Years' Experience in the African Methodist Episcopal Church* (Philadelphia: A.M.E. Bookrooms, 1881) 4–5. Wayman does not say what became of the wool after shearing. Perhaps it was for household use, although, as shown in the preceding chapter, this cannot be assumed; by the late 1820s many sheep farmers were supplying wool to factories.

2. "A. W. Wayman (Alexander Walker), 1821–1895," *Documenting the American South*, http://docsouth.unc.edu/church/wayman/bio.html; U.S. Census 1820, 198.

3. See, for example, Kenneth M. Stampp, *The Peculiar Institution: Slavery in the Ante-Bellum South* (New York: Vintage, 1956), 44–54; Eric Foner, *Reconstruction: America's Unfinished Revolution: 1863–1877* (New York: History Book Club, 2005) 126; and Baptist, *The Half Has Never Been Told*.

4. Judith A. Carney and Richard Nicholas Rosomoff, *In the Shadow of Slavery: Africa's Botanical Legacy in the New World* (Berkeley: University of California Press, 2009), 13–14; Richard C. Francis, *Domesticated: Evolution in a Manmade World* (New York: Norton, 2016), 166–167. One of Africa's endemic breeds, the fat-tailed sheep, was later imported to the United States. While most sources agree that sheep in Africa were mostly raised for meat, Salih Bilali, an enslaved African American Muslim in Georgia (born ca. 1765), recalled that in his West African homeland the Fulbe tribe raised sheep for wool. See Allan D. Austin, *African Muslims in Antebellum America: Transatlantic Stories and Spiritual Struggles* (New York: Routledge, 1997), 107.

5. Marcus Rediker, *The Slave Ship: A Human History* (New York: Penguin, 2007), 198.

6. PTJ, Retirement Series, 1:320–321. Spelling and punctuation modernized.

7. PTJ 37:331–332

8. See Baptist, *The Half Has Never Been Told*.

9. Charles Ball, *Slavery in the United States: A Narrative of the Life and Adventures of Charles Ball* (New York: Taylor, 1837), 146.

10. Louis Hughes, *Thirty Years a Slave: From Bondage to Freedom* (Milwaukee: South Side Printing, 1897), 31–32. See also Solomon Northup, *Twelve Years a Slave:*

Narrative of Solomon Northup (Auburn: Derby & Miller, 1853), 167. On the history of cotton culture in America and especially its effect on African Americans, see Baptist, *The Half Has Never Been Told*; Beckert, *Empire of Cotton*, 242–273; and Stephen Yafa, *Cotton: The Biography of a Revolutionary Fiber* (New York: Penguin, 2005), 147–172.

11. See, for example, Nicholas Herbemont, "Observations Suggested by the Late Occurrences in Charleston, by a Member of the Board of Public Works, of the State of Carolina," in *Pioneering American Wine: Writings of Nicholas Herbemont, Master Viticulturist*, edited by David S. Shields (Athens: University of Georgia Press, 2009), 257; Taylor, *Arator*, 248. This did not change with the end of the antebellum period; nearly twenty years after the Civil War, the commissioner of Georgia's Department of Agriculture cited low labor needs as a reason to increase the number of sheep in his state, writing, "The services of a single man will answer for the care of a thousand sheep, except at shearing time." J. C. Henderson, *A Manual of Sheep Husbandry in Georgia*, 2nd ed. (Atlanta: Georgia Department of Agriculture, 1883), 34.

12. PGW, Presidential Series, 12: 523–529.] Washington then lamented that he couldn't do so: "I know not the Negro among all mine, whose capacity, integrity, and attention could be relied on for such a trust as this." The statement that Washington had over five hundred sheep at all his properties is based on another letter of his of the same year: PGW, Presidential Series, 13, 6–13.

13. Randall, *Sheep Husbandry in the South*, 184–85.

14. Ibid., 61.

15. David Brion Davis, *Inhuman Bondage: The Rise and Fall of Slavery in the New World* (New York: Oxford University Press, 2006), 52.

16. Robert Somers, *The Southern States Since the War, 1870–1871* (London: Macmillan, 1871), 268.

17. JCC, 6:1080.

18. David S. Shields, introduction to Herbemont, *Pioneering American Wine*, 1, 9.

19. David Robertson, *Denmark Vesey: The Buried History of America's Largest Slave Rebellion and the Man Who Led It* (New York: Knopf, 1999), 4–5, 79–87.

20. Wilentz, *The Rise of American Democracy*, 237–239.

21. Robertson, *Denmark Vesey*, 4–5, 79–87.

22. Ibid., 6–7.

23. U.S. Census 1820, 115.

24. Herbemont's pamphlet, "Observations Suggested by the Late Occurrences in Charleston, by a Member of the Board of Public Works, of the State of Carolina," is reproduced in Herbemont, *Pioneering American Wine*, 251–260.

25. Ibid., 252.

26. Ibid., 252–53.

27. Ibid., 255.

28. James C. Bonner, *A History of Georgia Agriculture: 1732–1860* (Athens: University of Georgia Press, 1964) 13–17; Daniel J. Boorstin, *The Americans: The Colonial Experience* (New York: Random House, 1958), 83–84, 92–94.

29. Herbemont, *Pioneering American Wine*, 257.

30. U.S. Census 1840, 191.

31. U.S. Census 1860, *Agriculture of the United States*, 170.

32. U.S. Census 1860, 452.

33. E. Merton Coulter, *Daniel Lee, Agriculturist: His Life North and South* (Athens: University of Georgia Press, 1972), 77.

34. Ibid., 1–10, viii.

35. Ibid., 84–85. Unfortunately, I have been unable to locate copies of the *Augusta Weekly Constitutionalist* for the weeks in early 1859 in which Lee promoted revival of the transatlantic slave trade. I have had to rely on quotes and paraphrases Coulter includes in his biography of Lee. On efforts to resume the Atlantic slave trade just prior to the Civil War, see Don E. Fehrenbacher, *The Slaveholding Republic* (New York: Oxford University Press, 2001), 180–183.

36. U.S. Census, 1860, *Agriculture of the United States*, cix. The southern states together had just under five million sheep, up only about 435,000 head from the 1850 census. Ibid., cxxi.

37. Coulter, *Daniel Lee*, 84–85.

38. Davis, *Inhuman Bondage*, 175–192; Susan Dunn, *Dominion of Memories: Jefferson, Madison, and the Decline of Virginia* (New York: Basic Books, 2007), 49–60; Don E. Fehrenbacher, *The Dred Scott Case: Its Significance in American Law and Politics* (New York: Oxford University Press, 1978), 121.

39. William Peden, ed., *Notes of the State of Virginia* (Chapel Hill: University of North Carolina Press, 1982), 137–38.

40. St. George Tucker, *View of the Constitution of the United States with Selected Writings* (Indianapolis: Liberty Fund, 1999), viii–ix, 402–446.

41. Michael Kent Curtis, *No State Shall Abridge: The Fourteenth Amendment and the Bill of Rights* (Durham: Duke University Press, 1986), 30–31.

42. Ball, *Slavery in the United States,* 60.

43. On John Brown's sheep and their wool, see *Ohio Cultivator*, August 15, 1845; September 1, 1845; April 15, 1846; July 15, 1846; September 1, 1846; January 15, 1847; March 1, 1847; and May 15, 1847, in "Articles from The Ohio Cultivator on Perkins and Brown," West Virginia Archives and History, https://web.archive.org/web/20210427124148/http://www.wvculture.org/history/jbexhibit/ohiocultivator.html. See also David S. Reynolds, *John Brown, Abolitionist: The Man Who Killed Slavery, Sparked the Civil War, and Seeded Civil Rights* (New York: Knopf, 2005), 66–91. On the development of sheep culture in Ohio in the 1840's, see Stephen

L. Stover, "Early Sheep Husbandry in Ohio," *Agricultural History* 38, no. 2 (April 1962), 101–107.

44. Reynolds, *John Brown, Abolitionist*, 77.

45. Ibid., 80.

46. Ibid., 66.

47. Daegan Miller, *This Radical Land: A Natural History of American Dissent* (Chicago: University of Chicago Press, 2018), 79.

48. John Lord Hayes, trans., *Corolla Hymnorum Sacrorum: Being a Selection of Latin Hymns of the Early and Middle Ages* (Boston: Estes & Lauriat, 1887), 5–7.

49. Information is from Hayes's obituary in *Bulletin of the American Iron and Steel Association*, 21, no. 14, April 20 and 27, 1887, 105.

50. John L. Hayes, *Address before the National Association of Wool Manufacturers at the First Annual Meeting in Philadelphia, Sept 6, 1865* (Cambridge, Mass.: John Wilson, 1865).

51. John L. Hayes, "Sheep Husbandry in the South," in *Origin and Growth of Sheep Husbandry in the United States*, (Washington, D.C.: Government Printing Office, 1880), 65.

52. Ibid., 111. The same Senate document that contains Hayes's report has another unsigned report, also titled "Sheep Husbandry in the South," prepared by the Department of Agriculture. It mentions African Americans only in passing (e.g., on p. 49).

53. Ibid., 111.

54. Ibid., 111–112. A hostler was the person who cared for the horses of those staying at an inn.

55. Ibid., 112.

56. Paul S. Sutter, *Let Us Now Praise Famous Gullies: Providence Canyon and the Soils of the South* (Athens: University of Georgia Press, 2015), 152.

57. Herbemont, *Pioneering American Wine*, 255.

58. Randall, *Sheep Husbandry in the South*, 64. See also J. C. Henderson, *A Manual of Sheep Husbandry in Georgia* (Atlanta: Harrison, 1883), 25. "The herdsman is compelled to protect his flocks with gun or poison against the ravages of dogs."

59. "Simpson, Richard Franklin," *Biographical Directory of the United State Congress*, https://bioguide.congress.gov/search/bio/S000435.

60. Walter Lynwood Fleming, ed., *Documentary History of Reconstruction: Political, Military, Social, Religious, Educational and Industrial, 1865 to the Present Time*, vol. 1 (Cleveland: A. H. Clark, 1906), 280, 289–90; "Laws in Relation to Freedmen," 39th Cong., 2nd Sess., S. Doc. No. 6 (1867) at 195–196; Robert Cottrol and Raymond T. Diamond, "The Second Amendment: Toward an Afro-Americanist Reconsideration," *Georgetown Law Journal* 80, no. 309 (1991), 309–361, 344–45.

61. Alfred Avins, ed., *The Reconstruction Amendments' Debates* (Richmond: Virginia Commission on Constitutional Government, 1967), 209.

62. Charles Lane, *The Day Freedom Died: The Colfax Massacre, the Supreme Court, and the Betrayal of Reconstruction* (New York: Henry Holt, 2008), esp. 265–266. See also Lawrence Goldstone, *Inherently Unequal: The Betrayal of Equal Rights by the Supreme Court, 1865–1903* (New York: Walker, 2011), 88–98.

63. The counts are listed in the case report, United States v. Cruikshank, 92 U.S. 542, 545 (1876). Materials on the case, including the briefs of the federal government and other parties, are found in Philip B. Kurland and Gerhard Casper, *Landmark Briefs and Arguments of the Supreme Court of the United States: Constitutional Law*, vol. 7 (Arlington, Va.: University Publications of America, 1975), 287–314.

64. U.S. Fourteenth Amendment; United States v. Cruikshank; Goldstone, *Inherently Unequal*, 92–98; David P. Currie, *The Constitution in the Supreme Court: The First Hundred Years 1789–1888* (Chicago: University of Chicago Press, 1985), 395–396.

65. U.S. Constitution, Second Amendment; United States v. Cruikshank, 92 U.S. 542, 553 (1875).

66. See, for example, Cottrol and Diamond, "The Second Amendment," 342–349; and Stephen P. Halbrook, *Securing Civil Rights: Freedmen, the Fourteenth Amendment, and the Right to Bear Arms*, updated ed. (Oakland: Independent Institute, 2010).

67. Stephen J. Hahn, *The Roots of Southern Populism: Yeoman Farmers and the Transformation of the Georgia Upcountry 1850–1890*, (New York: Oxford University Press, 1983), 60–61, 239–268; Foner, *Reconstruction*, 203; J. Crawford King Jr., "The Closing of the Southern Range: An Exploratory Study," *Journal of Southern History* 48, no. 1 (February 1982): 53 (but see 55 for exceptions to the general practice).

68. Fleming, ed., *Documentary History of Reconstruction*, vol. 2, 27. See also King, "The Closing of the Southern Range."

69. King, "The Closing of the Southern Range," 57, although he notes that the introduction of barbed wire in 1879 reduced the need for timber.

70. Ibid., 59, 64, 65.

71. "Laws in Relation to Freedmen," 39th Cong., 2nd sess., S. Doc. No. 6, (1867) at 198.

72. Charles L. Flynn, *White Land, Black Labor: Caste and Class in Late Nineteenth-Century Georgia* (Baton Rouge: Louisiana State University Press, 1983), 128.

73. U.S. Census 1900, *Negroes in the United States*, 69.

74. Ibid., 73. Another bulletin affixed to the same census declares, "The 746,717 negro farmers, constituting 13 per cent of the farmers in the country . . . controlled

about 0.2 per cent of the total number of sheep in the country." U.S Census, 1900, *Agriculture, Part 1: Farms, Live Stock, and Animal Products*, ccix.

75. U.S. Census 1900, *Negroes in the United States*, 74, table 8, showing that 26.1 percent of these sheep were in the South Atlantic region (Delaware, Maryland, District of Columbia, Virginia, West Virginia, North Carolina, South Carolina, Georgia, and Florida) and 54.4 percent in the South Central region (Kentucky, Tennessee, Alabama, Mississippi, Louisiana, Texas, Oklahoma, and Arkansas).

76. Dan A. Rudd and Theodore Bond, *From Slavery to Wealth: The Life of Scott Bond* (Madison, Ark.: Journal Printing, 1917), 17, 362–63.

Chapter 5. Fido and the Fleece

1. Merrill F. Unger, *The New Unger's Bible Dictionary* (Chicago: Moody, 2005), 77, 1174.

2. All of the information presented here comes from the case report that came out of that appeal. See Miller v. the State, 5 Ga. App. 463, 63 S.E. 571, (1909).

3. Miller v. State, 464–65.

4. Ibid., 465–466.

5. Ibid., 465.

6. Ibid., 466. "Fice," meaning "a small dog," is an uncommon word in our time.

7. Daniel J. Boorstin, *The Mysterious Science of the Law: An Essay on Blackstone's Commentaries* (Chicago: University of Chicago Press, 1996), xiii–xvi; Norman F. Cantor, *Imagining the Law: Common Law and the Foundations of the American Legal System* (New York: Harper Perennial, 1996), 341–344; Gottfried Dietze, *The Federalist: A Classic on Federalism and Free Government* (Baltimore: Johns Hopkins University Press, 1999), 309–313.

8. *Blackstone's Commentaries on the Laws of England*, book 2, 389–390, Avalon Project, Yale Law School, avalon.law.yale.edu/subject_menus/blackstone.asp.

9. Ibid., 390, 393–94.

10. Keith Thomas, *Man and the Natural World: Changing Attitudes in England 1500–1800* (New York: Oxford University Press, 1983), 105.

11. Henry S. Randall, *Sheep Husbandry in the South* (Philadelphia: J. S. Skinner, 1848), 278–279.

12. PGW, Presidential Series, 10:484–488.

13. Coxe, *Tabular Statement of the Several Branches of American Manufactures*, table 46.

14. Richard Peters, "On Sheep Killing Dogs," *Memoirs of the Philadelphia Society for Promoting Agriculture* 2 (1811): 247–253.

15. Ibid, 251.

16. PGW, Presidential Series, 11: 519–524 (see also 625–631).

17. PTJ, 24:95–99.

18. PTJ, Retirement Series, 6:511–512.

19. PTJ, Retirement Series, 8:272.

20. PGW, Presidential Series, 11: 519–524.

21. "Thomas Jefferson to Edmund Bacon, 26 December 1808," NAF, https://founders.archives.gov/documents/Jefferson/99-01-02-9403.

22. Charles E. Beveridge and Charles Capen McLaughlin, eds., *The Papers of Frederick Law Olmsted: Slavery and the South, 1852–1857*, vol. 2 (Baltimore: Johns Hopkins University Press, 1981), 183. On Olmsted's time in the South, see Witold Rybczynski, *A Clearing in the Distance: Frederick Law Olmsted and America in the Nineteenth Century* (New York: Scribner, 1999), 109–131.

23. PTJ, Retirement Series, 1:15–19.

24. "Thomas Jefferson to James Ronaldson, 13 February 1809," NAF, https://founders.archives.gov/documents/Jefferson/99-01-02-9787.

25. PTJ, Retirement Series, 1:15–19.

26. Ibid.

27. PTJ, Retirement Series, 1:68–69.

28. *Acts of the General Assembly of the Commonwealth of Pennsylvania* (Philadelphia: John Bioren, 1809), 87–90.

29. *Acts of the General Assembly of the Commonwealth of Pennsylvania*, 89.

30. PTJ, Retirement Series, 4:161.

31. Ibid., 170.

32. Ibid., 346–349

33. Virginia Acts of Assembly, 1813–1814, 55. By calling sheep improvement "important . . . especially at this time," the Virginia legislature appears to have been noting the effect of the War of 1812 on wool supplies.

34. Ibid., 55–56.

35. Hayes, *Origin and Growth of Sheep Husbandry in the United States*, 33–34.

36. Ibid., 32. The word "cur"—like "fice," noted earlier—is uncommon in our time. It describes an aggressive dog, especially one that is a mutt.

37. Ibid., 33.

38. For the Hereford weights, see "Hereford Cattle," *Breeds of Livestock*, Oklahoma State University, http://afs.okstate.edu/breeds/cattle/hereford/index-2.html.

39. Hayes, *Origin and Growth of Sheep Husbandry in the United States*, 35. Denny is described as being from "Boyle County, Missouri." This, however, must be a misprint; there is not now nor was there such a county in that state. Furthermore, as noted below Missouri *did* enact a dog tax in 1877, thus it would be peculiar for someone to complain that his state did not have such a tax. Probably Denny was actually from Boyle County, Kentucky.

40. For the discussion of *Burden v. Hornsby*, I have relied on Walter L. Chaney, "The True Story of 'Old Drum,'" *Missouri Historical Review* 19, no. 2 (January 1925): 313–324; Edwin M. C. French, *Senator Vest: Champion of the Dog* (Boston: Meador, 1930); Icie F. Johnson, *The Old Drum Story* (Warrensburg, Mo.: Chamber of Commerce, 1957); George Ganter, "Old Drum: Immortal Missouri Hound," *Journal of the Missouri Bar* 64 (2008) 138–140 (reprint of article first appearing in the same periodical in April 1940); and "Man's Best Friend: The Old Drum Story," Missouri State Archives, https://www.sos.mo.gov/archives/education/olddrum/StoryofBurdenvHornsby.asp, which contains links to transcripts of some of the original documents. While the verdict in *Burden v. Hornsby* was appealed to the Missouri Supreme Court—which affirmed the decision—the appeal was strictly over procedural matters that are not relevant to the discussion here (see *Burden v. Hornsby*, 50 Mo. Reports 238 [1872]). Thus there is no official report considering the merits of the case.

41. In describing Dick Ferguson as a young man, I am following the description of him in Chaney, "The True Story of Old Drum," 315. Accounts on Ferguson vary, however. In Johnson, *The Old Drum Story*, Ferguson is called "a young boy Mr. Hornsby was rearing," while in Ganter, "Old Drum," 139, a reference to Hornsby refers to Ferguson as "his man." As was the case with Mr. Miller's son in Georgia, it is not clear how old Ferguson was, only that he was acting on the instructions of an adult property owner who wanted trespassing dogs shot because they were killing his sheep. It seems logical to prefer Chaney's portrayal of Ferguson since he gives the most thorough description of the Hornsby household.

42. U.S. Census 1870, *The Statistics of Wealth and Industry in the United States*, 189.

43. This information was obtained from the archives of the State Historical Society of Missouri, listed under "1870 Agricultural Census for Johnson County."

44. The most thorough description of the evidence and the testimony is in Chaney, "The True Story of 'Old Drum,'" 319–321.

45. French, *Senator Vest*, 47.

46. Ibid., 47–58; Douglas Brinkley, *The Wilderness Warrior: Theodore Roosevelt and the Crusade for America* (New York: Harper Collins, 2009), 206, 240.

47. French, *Senator Vest*, 21–22; Joseph Tartakovsky, *The Lives of the Constitution: Ten Exceptional Minds That Shaped America's Supreme Law* (New York: Encounter, 2018), 151–152.

48. For the entire three-paragraph speech, see "Eulogy of the Dog," U.S. Senate, https://www.senate.gov/artandhistory/history/resources/pdf/VestDog.pdf.

49. Johnson, *The Old Drum Story*, 8–9.

50. French, *Senator Vest*, 32–33.

51. Ganter, "Old Drum," 138.

52. Missouri State Archives, https://www.sos.mo.gov/CMSImages/MDH

/TranscriptofProceedingsbeforeJustieceofPeace,February5,1870.pdf.

53. This information was obtained from the archives of the State Historical Society of Missouri, listed under "1870 Agricultural Census for Johnson County."

54. Missouri State Archives, https://www.sos.mo.gov/CMSImages/MDH/TranscriptofProceedingsbeforeJustieceofPeace,February5,1870.pdf.

55. John Gardner, *Grendel* (New York: Knopf, 1971). The book begins with Grendel describing an encounter he had with a ram.

56. Miller v. State, 464.

57. *Laws of Missouri Passed at the Regular Session of the Twenty-Ninth General Assembly* (Jefferson City, Mo.: Regan & Carter, 1877), 326. Another part of the law was, of course, a tax on dogs; section 3 assessed a fee of one dollar for every canine kept.

58. Phillips v. Lewis, 3 Tenn. Cases 230 (1877), 252–253. See also Chandler v. the State, 69 Tenn. Reports 296 (1878).

59. Samuel W. Small, *A Stenographic Report of the Proceedings of the Constitutional Convention Held in Atlanta, Georgia, 1877* (Atlanta: Constitution, 1877), 29.

60. 1877 Georgia Constitution, article 7, section 2, paragraph 1.

61. Official Code of Georgia, § 4-8-5, "Cruelty to dogs; authorized killing of dogs" (2018). See also § 4-8-4, "Liability of owner or custodian for damages done to livestock, poultry, or pet animal by dog."

62. Official Code of Georgia § 4-3-2 (2018).

63. New Jersey Statutes 4, 19–9, "Right to Destroy Offending Dogs" (2018).

Chapter 6. Mary Had a Little Lamb

1. David Hackett Fischer, *Albion's Seed: Four British Folkways in America* (New York: Oxford University Press, 1989), 94.

2. Ruth E. Finley, *The Lady of Godey's: Sarah Josepha Hale* (Philadelphia: J. B. Lippincott, 1931), 279.

3. Iona and Peter Opie, eds., *The Oxford Dictionary of Nursery Rhymes* (Oxford: Oxford University Press, 1951), 300. Note, however, that "Mary's Lamb" as first published consisted of three eight-line stanzas, as opposed to the four stanzas of four lines Lucas describes. See Sarah Josepha Hale, *Poems for Our Children* (Boston: Marsh, Capen & Lyon, 1830), 6–7. The four-stanza version is the one commonly anthologized; it is the one reprinted in *The Oxford Dictionary of Nursery Rhymes* and on the Poetry Foundation's website: https://www.poetryfoundation.org/poems/44322/marys-lamb.

4. Sherbrooke Rogers, *Sarah Josepha Hale: A New England Pioneer* (Grantham, N.H.: Tompson & Rutter, 1985), 14.

5. Finley, *The Lady of Godey's*, 289. See also Richard Walden Hale, "Mary Had a Little Lamb and Its Author," *Century Illustrated Monthly Magazine*, March 1904, 738–742.

6. Except where noted, the information here and in the next four paragraphs is from Sarah Josepha Hale's autobiographical sketch in a book of poems she edited, *The Ladies Wreath* (Boston: Marsh, Capen & Lyon, 1837), 383–388.

7. Peggy M. Baker, *The Godmother of Thanksgiving: The Story of Sarah Josepha Hale* (Plymouth, Mass.: Pilgrim Society & Pilgrim Hall Museum, 2007), 4, https://web.archive.org/web/20160328171341/http://www.pilgrimhallmuseum.org/pdf/Godmother_of_Thanksgiving.pdf.

8. Ibid., 7–15.

9. Hale, *Poems for Our Children*.

10. Steven Watts, *The People's Tycoon: Henry Ford and the American Century* (New York: Alfred E. Knopf, 2005), 10–13, 403. The characterization of Ford as possibly the most famous man in the world in the 1920s is also from *The People's Tycoon*, 3.

11. Charles C. Calhoun, *Longfellow: A Rediscovered Life* (Boston: Beacon Press, 2004), 190, 198.

12. Henry Wadsworth Longfellow, *Tales of a Wayside Inn* (Boston: Ticknor & Fields, 1863), 18–25.

13. Watts, *The People's Tycoon*, 404; Allan Nevins and Frank Ernest Hill, *Ford: Expansion and Challenge, 1915–1933* (New York: Scribner, 1957), 498–499.

14. Watts, *The People's Tycoon*, 404.

15. Nevins and Hill, *Ford*, 499; *The Story of Mary's Little Lamb* (Sudbury, Mass.: Ford, 1928), 1–2.

16. For this account of Tyler's story I have relied on *The Story of Mary's Little Lamb*, 4–10; Finley, *The Lady of Godey's*, 285–289; and Hale, "Mary Had a Little Lamb and Its Author," 738–742.

17. Finley, *The Lady of Godey's*, 287. Tyler maintained that she had lost the paper over the years. See Hale, "Mary Had a Little Lamb and Its Author," 739.

18. Hale, "Mary Had a Little Lamb and Its Author," 742.

19. Hale, *Poems for Our Children* (Boston: Marsh, Capen & Lyon, 1916), 4.

20. *The Story of Mary's Little Lamb*, 13.

21. Finley, *The Lady of Godey's*, 301.

22. Ibid., 282–283.

23. "'Twas Lucy Had the Little Lamb—Says Nagle's Old-Time School Book," *Boston Daily Globe*, January 20, 1927, 19. The book was titled *Town's Second Reader* (Portland: Sanborn & Carter, 1851).

24. Finley, *The Lady of Godey's*, 292.

25. "Ford Now Retracts Attacks on Jews; Orders Them Ended," *New York Times*,

July 8, 1927, 1; "Ford and Sapiro Settle Libel Suit," *New York Times*, July 17, 1927, 1. Ruth Finley in *The Lady of Godey's* reprints a few lines from Ford's apology (202), but the lines she quotes are general and do not indicate that this was an expression of regret for other articles that were anti-Semitic.

26. "Ford Now Retracts Attacks on Jews."

27. The controversy is not covered in Watts, *The People's Tycoon*, or in Richard Snow, *I Invented the Modern Age: The Rise of Henry Ford* (New York: Scribner, 2013). One biography that does cover the matter, albeit briefly, is Nevins and Hill, *Ford*, 499.

28. Michael R. Canfield, *Theodore Roosevelt in the Field* (Chicago: University of Chicago Press, 2015), 98–103; Darrin Lunde, *The Naturalist: Theodore Roosevelt, a Lifetime of Exploration, and the Triumph of American Natural History* (New York: Crown, 2016), 143–149.

29. "Badlands trilogy" is from Clay S. Jenkinson, "The Cowboy, the Crusader, and the Salvation of the American Buffalo," in *Theodore Roosevelt: Naturalist in the Arena*, ed. Char Miller and Clay S. Jenkinson (Lincoln: University of Nebraska Press, 2020), 144.

30. Theodore Roosevelt, *Hunting Trips of a Ranchman: Sketches of Sport on the Northern Cattle Plains; and The Wilderness Hunter: An Account of the Big Game of the United States and Its Chase with Horse, Hound, and Rifle* (New York: Modern Library, 1996), 121.

31. Ibid., 105. See also Roosevelt, *The Wilderness Hunter*, 480, 778.

32. Don Rickey Jr., *Cowboy Dress, Arms Tools and Equipments, as Used in the Little Missouri Range Country and the Medora area, in the 1880's* (1957), National Park Service, npshistory.com/publications/thro/rickey.pdf. The Little Bighorn Battlefield National Monument was called the Custer Battlefield National Monument when Rickey wrote this report.

33. Ibid., references to wool clothing at 4, 6, 8, 9–11, 17, 21, and 25.

34. U.S. Census 1880, *Report on the Productions of Agriculture*, 5.

35. Ibid., vii.

36. Clarence W. Gordon, *Report on Cattle, Sheep, and Swine, Supplementary to Enumeration of Live Stock on Farms in 1880* (Washington, D.C.: Government Printing Office, 1883), hereafter *Gordon's Report*. The document included statistics from Florida, which was considered to have similar characteristics with the West. The Florida statistics are eliminated from this analysis.

37. Ibid., 149.

38. Ibid., 53.

39. Ibid., 51, 54.

40. Brinkley, *The Wilderness Warrior*, 196–198; Thomas Bailey and Katherine Joslin, *Theodore Roosevelt: A Literary Life* (Lebanon, N.H.: ForeEdge, 2018), 75–76.

41. Stephen Ambrose, introduction to Roosevelt, *Hunting Trips of a Ranchman*, xxi.

42. *Gordon's Report*, 53.

43. Ibid., 73.

44. Ibid., 81–82.

45. Ibid., 91.

46. Theodore Roosevelt, *The Winning of the West*, vol. 3 (New York: Putnam, 1894), 15.

47. Oscar Wilde, *The Picture of Dorian Gray* (Oxford: Oxford University Press, 2006), ix, 6.

48. For more on the exposition, see Stanley Appelbaum, *The Chicago World's Fair of 1893: A Photographic Record* (New York: Dover, 1980); Neil Harris, Wim de Wit, James Gilbert, and Robert W. Rydell, *Grand Illusions: Chicago's World's Fair of 1893* (Chicago: Chicago Historical Society, 1993); and Norman Bolotin and Christine Laing, *The World's Columbian Exposition: The Chicago World's Fair of 1893* (Urbana: University of Illinois Press, 1992).

49. Appelbaum, *The Chicago World's Fair of 1893*, 106; Wim de Wit, "Building an Illusion: The Design of the World's Columbian Exposition," in Harris et al., *Grand Illusions*, 96n2.

50. Martin Ridge, "The Life of an Idea: The Significance of Frederick Jackson Turner's Frontier Thesis," *Montana: The Magazine of Western History*, (Winter 1991): 7.

51. "Turner (Frederick Jackson Papers)," Online Archive of California, https://oac.cdlib.org/findaid/ark:/13030/tf4r29n77v.

52. Gene M. Gressley, "The Turner Thesis: A Problem in Historiography," *Agricultural History* 32, no. 4 (October 1958): 227; Ridge, "The Life of an Idea," 3.

53. U.S. Census 1890, xxxiv. The official census report was not published until 1895, so clearly Turner must have read a preliminary bulletin issued prior to the final document.

54. Frederick Jackson Turner, "The Significance of the Frontier in American History," in *The Frontier in American History* (1920; repr., New York: Dover, 1996), 1.

55. Ibid., 11–12.

56. U.S. Census 1870, *The Statistics of the Wealth and Industry of the United States*, 74, 75.

57. Turner, "The Significance of the Frontier in American History," 12, 16, and 18.

58. Frederick Jackson Turner, "The Middle West" in *The Frontier in American History*, 126–156. The essay appeared in the December 1901 issue of *International Monthly*.

59. Ibid., 144, 147, 151.

60. Ibid., 151.

61. Ibid., 129, 143, 151.

62. Biographical details of Wister are from James Fleming Hosic, introduction to Owen Wister, *The Virginian* (New York: Macmillan Company, 1917), ix–xi.

63. At a dinner with acquaintances in 1891, Wister "wondered aloud why no one had yet featured the American West in the way that Rudyard Kipling was using India as a setting for fictional tales." See Christopher Knowlton, *Cattle Kingdom: The Hidden History of the Cowboy West* (Boston: Houghton Mifflin Harcourt, 2017), 335.

64. Ibid., x.

65. Ibid., xiii.

66. U.S. Census 1880, *Report on the Statistics of Agriculture*, 141; U.S. Census 1890, *Report on the Statistics of Agriculture*, 29, 236, 274. The census figures for Wyoming sheep in 1890 counted only farm animals; there was no tabulation of range sheep as there had been in 1880. In neighboring Montana, however, the number of range sheep had grown from 95,000 in 1880 to almost 494,000 in 1890, a 420 percent increase. If between 1880 and 1890 the increase of sheep in Wyoming was only half that, a 210 percent increase, then the state would have had over 650,000 range sheep in addition to the 712,000 reported on farms. The Wyoming sheep count in 1890 would thus have stood at well over 1.3 million.

67. Wister, *The Virginian*, 157, 71, 356.

68. Ibid., 43, 388. In the second reference a bighorn is called a "ram," the only use of that word in the book. There are also two uses of "lamb," but the word is employed as a metaphorical description of people (101, 205).

69. Knowlton, *Cattle Kingdom*, 336.

70. Philip Durham and Everett L. Jones, *The Negro Cowboys* (Lincoln: University of Nebraska Press, 1965), 225. Given the focus of their book, Durham and Jones note the lack of African Americans in *The Virginian*. Like the sheep, black men and women didn't appear in Wister's novel, even though they were an integral part of the West.

71. "The Big Valley," IMDb, https://www.imdb.com/title/tt0058791/?ref_=nv_sr_1?ref_=nv_sr_1.

72. U.S. Census 1880, *Report on the Productions of Agriculture*, 144. Gordon's report was not broken down by county, so it is not known how many range sheep or cattle there were in San Joaquin County besides those on farms, but clearly overall this was sheep country. A real-life Barkley family would have been far more likely to have prospered from raising sheep than from raising cattle.

73. Joe B. Frantz and Julian Ernest Choate Jr., *The American Cowboy: The Myth and the Reality* (Norman: University of Oklahoma Press, 1955), 113–114. On the Pleasant Valley War, see Daniel Justin Herman, *Hell on the Range: A Story of Honor, Conscience, and the American West* (New Haven: Yale University Press, 2010); and Eduardo Obregón Pagán, *Valley of the Guns: The Pleasant Valley War and the Trauma*

of Violence, (Norman: University of Oklahoma Press, 2018).

74. Upton Sinclair, *The Jungle* (New York: Grosset & Dunlap, 1906), 38.

75. Theodore Roosevelt, "The Northwest in the Nation," *Proceedings of the State Historical Society of Wisconsin at Its Fortieth Annual Meeting* (Madison: Democrat Printing Company, 1893), 92–99.

76. Ibid, 97.

77. Ibid.

78. Brinkley, *The Wilderness Warrior*, 241.

79. Ibid, 241. See also Bailey and Joslin, *Theodore Roosevelt*, 88–90.

80. Brinkley, *The Wilderness Warrior*, 463–464; Bailey and Joslin, *Theodore Roosevelt*, 138–139.

81. Turner, "The Significance of the Frontier in American History," 15.

82. Wister, *The Virginian*, 291.

83. Roosevelt, *The Wilderness Hunter*, in *Hunting Trips of a Ranchman*, 762. On the importance of manliness to Roosevelt, see Gail Bederman, *Manliness and Civilization* (Chicago: University of Chicago Press, 1995), 170–215.

84. Roosevelt, *The Wilderness Hunter*, 365.

85. Ibid., 429.

86. Ibid., 329.

87. In Roosevelt's case, his speeches also have been described as having "an aggressively masculine style." Kenneth Cmiel, *Democratic Eloquence: The Fight over Popular Speech in Nineteenth-Century America* (New York: William Morrow, 1990), 250.

88. Frantz and Choate, *The American Cowboy*, 113. Herman, *Hell on the Range*, 103.

89. Glenda Riley, *The Female Frontier: A Comparative View of Women on the Prairie and the Plains* (Lawrence: University Press of Kansas, 1988), 1. Wister's *The Virginian* has come under similar criticism, with one critic writing that there is a "presumed superiority of men and their dominion over women: the women in the book are either whores or saints whose salvation lies in submitting to the more powerful men." Quoted in Knowlton, *Cattle Kingdom*, 338.

90. Ibid., 21, 117, 184.

91. Ibid., 201.

92. U.S. Census 1910, *Agriculture: General Report and Analytical Tables*, 625.

93. Ibid., 626.

94. The poster is in the National Archives labeled "Food Administration—Posters—Shepherdesses drive flock through crowded Chicago rousing interest in sheep raising campaign": https://catalog.archives.gov/id/31481135.

Chapter 7. Machines in the Pasture

1. The Rural Electrification Act, authorizing federal government loans for the construction and maintenance of the infrastructure necessary to bring electricity to rural areas, was passed in 1936. See U.S. Stats., 49:1363.

2. Annie Tindley and Andrew Wodehouse, *Design, Technology, and Communication in the British Empire, 1830–1914* (London: Palgrave Macmillan, 2016) 20, 79; Evan McHugh, *The Shearers: The Story of Australia as Told from the Woolsheds* (Melbourne: Penguin, 2015) 30–36. The Wolseley Company remains in operation. Over the years there were a number of mergers and acquisitions, and in 2017 it changed its name to Ferguson PLC. In the United States its chief business is as a distributor of plumbing and heating products. See "About Us," Ferguson, https://www.corporate.ferguson.com/about-us/about-us/default.aspx.

3. Stuart Svenson, *The Shearers' War: The Story of the 1891 Shearers' Strike* (St. Lucia, Australia: University of Queensland Press, 1989), 21.

4. E. S. Bartlett, "Shearing," in *The Golden Hoof: A Practical Sheep Book*, P. V. Ewing, ed. (Chicago: Sheep Breeder, 1936), 97. Over a decade later, sheepmen in the western United States still spoke of taking their sheep to an "Australian shearing plant" to be sheared. See U.S. Tariff Commission, *The Wool-Growing Industry* (Washington, D.C.: Government Printing Office, 1921), 487, 489.

5. Archer B. Gilfillan, *Sheep: Life on the South Dakota Range* (1929; repr., St. Paul: Minnesota Historical Society, 1993), 133. (Reprint of original 1929 edition.)

6. Hughie Call, *Golden Fleece* (Lincoln: University of Nebraska Press, 1981), 151.

7. Ibid, 151. See also Gilfillan, *Sheep*, 133; U.S. Tariff Commission, *The Wool-Growing Industry*, 529. Complaints about sheep being harmed by too-close shearing were raised well before Call and Gilfillan chronicled their experiences. In 1914 the Massachusetts Society for the Prevention of Cruelty to Animals and other organizations appealed to the governor of Texas, protesting that sheep there were being sheared too closely and shorn in cold weather, leading to their deaths from exposure. The humane organizations urged Texas lawmakers "that such cruelty be branded as a misdemeanor and suitable punishment fixed for its practice." See "Want Sheep Protected during Shearing Season," *Austin Daily Statesman*, March 17, 1914, 10.

8. Svenson, *The Shearers' War*, 21. Svenson further writes that machine shearing gave an average yield of eight ounces more per sheep than hand clipping, in essence arguing that machines were superior for both sheep safety and human economy.

9. L. A. Morrell, *The American Shepherd: Being a History of the Sheep* (New York: Harper, 1863), 177–178.

10. It seems that the shearer's skill is more significant than whether his tool is mechanized or not, even considering the assertion Hughie Call (and others) made

that shearing is faster when done by machine than by hand. The famed Australian shearer Jack Howe in 1892 set a record by hand shearing 321 sheep in seven hours and forty minutes. Later that year he set a new record by machine shearing 237 sheep in a day—in other words, he was far more productive with manual tools than with a machine. See McHugh, *The Shearers*, 96–97.

11. Svenson, *The Shearers' War*, 45–46.

12. Gilfillan, *Sheep*, 132–133; Call, *Golden Fleece*, 148.

13. McHugh, *The Shearers*, 242–243; Kerry McGawley, "An Analysis of Data from Dwayne Black's 9-Hour World Record 05-04-05," Blackpowered.com, www.shearingworld.com/blackpowered/datauwa.htm; "Study Shows Shearing Is Toughest Job," *World Today*, ABC (Australian Broadcasting Corporation), February 25, 2000.

14. U.S. Tariff Commission, *The Wool-Growing Industry*, 178; Call, *Golden Fleece*, 150.

15. U.S. Tariff Commission, *The Wool-Growing Industry*, 178–179.

16. *Casper (WY) Daily Tribune*, June 6, 1918, 1.

17. "Shearers Strike Cut Short at Cadoma at Warning of Sheriff," *Casper (WY) Daily Tribune*, June 6, 1918, 1.

18. "Strike of Sheep Shearers Will Lead to Federal Prosecution if Wool Clip Is Delayed, Says Houx," *Casper (WY) Daily Tribune*, May 21, 1918, 1. The article miscalculates seventeen cents per sheep as amounting to thirty dollars per day.

19. Ibid.

20. In March 1917, the U.S. government consumed seventy-three million pounds of wool for the war effort, to that point the largest one-month consumption of wool in American history. The secretary of the National Woolgrowers' Association reported that this wool was used to make "low-grade goods, largely for army purposes, [that] were one-fourth or one-third heavier than the goods ordinarily required by the civilian population." See U.S. Tariff Commission, *The Wool-Growing Industry*, 526.

21. U.S. Tariff Commission, *The Wool-Growing Industry*, 77.

22. "Strike of Shearers at Opal Compromised," *Casper (WY) Daily Tribune*, April 22, 1926, 1.

23. "Sheep Shearers Strike Seen as Threat to Agricultural Industry in California," *Petaluma (CA) Argus-Courier*, March 30, 1938, 7; "Sheep Shearers Strike Will Probably Be a Fore-runner of Invasion of Fall Harvests," *Ukiah (CA) Republican Press*, April 13, 1938, 7. The Sheep Shearers Union of North America was founded in Butte, Montana, in 1903, which helps explain why the California strike received extensive coverage in Montana papers. See Pat Hansen, "Sheep Shearers Union/Merchandise and Commission Co.," *Montana Standard*, May 1, 2002.

24. "Nationwide Shearers Strike Made Effective," *Great Falls (MT) Tribune*, April 10, 1938, 2.

25. "Police Protect Sheep Shearers," *Oakland Tribune*, April 18, 1938, 4.

26. "Sheep Shearers Return to Jobs," *San Francisco Examiner*, April 22, 1938, 7. The strike was not officially abandoned until July the following year. See "Sheep Shearers Vote for New Affiliation," *Helena (MT) Daily Independent*, July 19, 1939, 3.

27. U.S. Tariff Commission, *The Wool-Growing Industry*, 178, 477.

28. Ibid., 79; Gavin Weightman, *The Industrial Revolutionaries: The Making of the Modern World, 1776–1914* (New York: Grove, 2007), 314–318; "Our History," Wolseley Company, https://web.archive.org/web/20210325154002/http://corporate.wolseley.co.uk/about-us/our-history.aspx.

29. The film is in the National Archives and can be viewed online: *Wool Processing, Shearing Sheep*, https://catalog.archives.gov/id/91950.

30. A second film, with the same title, is similar; it has no sheep but shows wool being bundled, processed into cloth, and fashioned into upholstery for Ford vehicles. The caption accompanying this film reads "'From raw materials to finished product' is a phrase applied to many activities of the Ford Motor Company." The film is in the National Archives and can be viewed online: *Wool Processing, Shearing Sheep*, https://catalog.archives.gov/id/91951. The National Archives lists the date of these films as "ca. 1926;" if either of them were actually made after 1927, the wool would have wound up in not in a Model T but in a Model A instead. Ford manufactured the Model T from 1908 through 1927, replacing it then with the Model A. See "Company Timeline," Ford Motor Company, https://corporate.ford.com/history.html.

31. "Wool Display, Ford Exposition, New York World's Fair, 1939," Henry Ford Museum of American Innovation, https://www.thehenryford.org/collections-and-research/digital-collections/artifact/373186 and https://www.thehenryford.org/collections-and-research/digital-collections/artifact/373190.

32. "Wool Display, Ford Exposition, New York World's Fair, 1939," Henry Ford Museum of American Innovation, https://www.thehenryford.org/collections-and-research/digital-collections/artifact/373186.

33. Hayes, *Origin and Growth of Sheep Husbandry in the United States*; George Fayette Thompson, *Information Concerning the Angora Goat* (Washington, D.C.: Government Printing Office, 1901), 12, 18; Nellie Peters Black, *Richard Peters: His Ancestors and Descendants 1810–1889* (Atlanta: Foote & Davies, 1904), 41–42.

34. An early catalog of Model T accessories contained a listing for wool gaskets "made of best quality wool felt in the correct shapes and dimensions to fit the parts specified." Floyd Clymer, *Henry's Wonderful Model T 1908–1927* (New York: McGraw Hill, 1955), 208.

35. Frederick Law Olmsted, *The Papers of Frederick Law Olmsted*, edited by Charles E. Beveridge et al. (Baltimore: Johns Hopkins University Press, 1977–2015), 8:264; Rybczynski, *A Clearing in the Distance*, 360–364; and Charles E. Beveridge, "Olmsted: His Essential Theory," *Nineteenth Century* 20, no. 2 (Fall 2000): 32–37.

36. Clive Gravett, *Two Men Went to Mow: The Obsession, Impact and History of Lawn Mowing* (London: Unicorn, 2018), 22–33, 41, 55; Virginia Scott Jenkins, *The Lawn: A History of an American Obsession* (Washington, D.C.: Smithsonian, 1994), 29–30. Early mowers were manually operated devices pushed by human hands or pulled by draft animals; engine-powered mowers did not appear until early in the twentieth century.

37. Gravett, *Two Men Went to Mow*, 10; Hedrick, *A History of Horticulture*, 263.

38. PTJ, 9:369–375.

39. Gravett, *Two Men Went to Mow*, 18–19; also Jenkins, *The Lawn*, 10; Ulysses Prentiss Hedrick, *A History of Horticulture in America to 1860* (New York: Oxford University Press, 1950), 263.

40. *Annual Report of the Board of Commissioners of the Central Park* (New York: Bryant), hereafter *Ann. Rep. Central Park* 10 (1867), 16. Presumably the Central Park mowers were British imports; the first patent in the United States for a lawn mower was issued in 1868, nearly forty years after Budding's British patent. See M. D. Leggett, *Subject Matter Index of Patents for Inventors Issued by the United States Patent Office From 1790 to 1873, Inclusive*, vol. 2 (Washington, D.C.: Government Printing Office, 1874), 949.

41. *Ann. Rep. Central Park* 10 (1867), 17.

42. Ibid., 64. Dollar calculations from Inflation Calculator, Official Data Foundation, https://www.officialdata.org/.

43. Olmsted, *The Papers of Frederick Law Olmsted*, 3:126.

44. *Ann. Rep. Central Park* 8 (1865), 39–42.

45. *Ann. Rep. Central Park* 9 (1866), 43 (1866).

46. *Ann. Rep. Central Park* 6 (1863), 27–28.

47. Roy Rosenzweig and Elizabeth Blackmar, *The Park and the People: A History of Central Park* (Ithaca, N.Y.: Cornell University Press, 1992), 252; Justin Martin, *Genius of Place: The Life of Frederick Law Olmsted* (Boston: Da Capo, 2011), 235; "Sheep Meadow," Central Park Conservancy http://www.centralparknyc.org/things-to-see-and-do/attractions/sheep-meadow.html.

48. Martin, *Genius of Place*, 235. See also Rosenzweig and Blackmar, *The Park and the People*, 252.

49. *Ann. Rep. Central Park* 6 (1863), 5–7.

50. *Ann. Rep. Central Park* 4 (1861), 106–107.

51. Olmsted, *The Papers of Frederick Law Olmsted*, 3:78; Rybczynski, *A Clearing in the Distance*, 155–157.

52. Frederick Law Olmsted, *The Spoils of the Park* (F. Olmsted, 1882), 22.

53. Olmsted, *The Spoils of the Park*, 55.

54. Barbara McEwan, *White House Landscapes: Horticultural Achievements of American Presidents* (New York: Walker, 1992) 85, 106, 142; William Seale, *The White House Garden* (Washington, D.C.: White House Historical Association, 1970) 27, 106.

55. "Sheep to Graze at White House," *Washington Post*, April 30, 1918, 2. See also "Why Did President Woodrow Wilson Keep a Flock of Sheep on the White House Lawn?," White House Historical Association, https://web.archive.org/web/20170110233518/https://www.whitehousehistory.org/questions/why-did-president-woodrow-wilson-keep-a-flock-of-sheep-on-the-white-house-lawn; and Brian Resnick and National Journal, "White House Sheep: A History," *Atlantic*, October 17, 2014, https://www.theatlantic.com/politics/archive/2014/10/white-house-sheep-a-history/453405.

56. "White House Sheep 'Baa,' Have to be 'Shooed' Away, *Washington Post*, August 20, 1919, 1.

57. "Wool of White House Sheep Is to Be Sold to Aid Red Cross Fund, *Washington Post*, May 18, 1918, 2; "More White House Sheep," *Washington Post*, March 3, 1919, 9. For some reason, there later appeared reports that the 1918 wool sale raised over $52,000, considerably more than the original figure relayed by the press. It is possible that the higher sum was from a 1919 shearing, when the flock had increased and so provided more wool, or that the $52,000 represented the combined auction sales from both 1918 and 1919. See "President to Sell His Sheep," *Nebraska State Journal*, August 1, 1920, 1. Equivalent value of $30,000 in 2020 calculated from Inflation Calculator, Official Data Foundation, https://www.officialdata.org/.

58. "Klondike Asks Job as Harding's Shepherd of White House Flock," *Washington Post*, July 17, 1920, 1; "President Going out of Sheep Business," *Philadelphia Inquirer*, August 1, 1920, 1.

59. *Harper's Weekly*, September 30, 1899, 991.

60. U.S. Tariff Commission, *The Wool-Growing Industry*, 557.

61. Ibid., 591. The market for lamb and mutton was said to have been expanding slowly over the preceding 150 years. Ibid., 259.

62. Ibid., 591. The figure of 742 million pounds of lamb and mutton consumed is based on the 1920 U.S. Census, which counted slightly over 106 million people.

63. Ibid., 466, 473, 491, 497, 501, 504, 516, 533, 548, 551, 568, 582, 589.

64. Ibid., 591. See also John P. Huttman, "British Meat Imports in the Free Trade Era," *Agricultural History* 52, no. 2 (April 1978): 247–262, 250; Woods, *The Herds Shot Round the*, 118.

65. Sam Bowers Hilliard, *Hog Meat and Hoecake: Food Supply in the Old South, 1840–1860* (Athens: University of Georgia Press, 2014), 142.

66. Henry S. Randall, *Sheep Husbandry in the South* (Philadelphia: J. S. Skinner, 1848), 56. See also Wright, *Wool-Growing and the Tariff*, 125.

67. JCC, 1:78.

68. Randall, *Sheep Husbandry in the South*, 137, 158; Woods, *The Herds Shot Round the World*, 52–77.

69. Wright, *Wool-Growing and the Tariff*, 66.

70. Mary Randolph, *The Virginia House-wife*, (1824; repr., Columbia: University of South Carolina Press, 1984), ix, 56–62.

71. It was different with hogs because their meat could be preserved through salting; thus pork could remain edible well past slaughtering. See Mark Essig, *Lesser Beasts: A Snout to Tail History of the Humble Pig* (New York: Basic Books, 2015), 85–86, 177–178.

72. Carl Sandburg, "Chicago," in *Chicago Poems* (New York: Holt, 1916), 3–4.

73. W. Joseph Grand, *Illustrated History of the Union Stockyards* (Chicago: Thomas Knapp, 1896), 8.

74. Ibid., 9–11; Dominic A. Pacyga, *Slaughterhouse: Chicago's Union Stock Yard and the World It Made* (Chicago: The University of Chicago Press, 2015), 29–61; "The Birth of the Chicago Union Stockyards," Chicago Historical Association, https://web.archive.org/web/20110319025045/http://www.chicagohs.org:80/history/stockyard/stock1.html.

75. Grand, *Illustrated History of the Union Stockyards*, 18.

76. Ibid., 13.

77. Pacyga, *Slaughterhouse*, 15. In the late nineteenth and early twentieth centuries these processing plants were a prominent tourist attraction. See Pacyga, 5–27.

78. Woods, *The Herds Shot Round the World*, 119–120; William J. Bernstein, *A Splendid Exchange: How Trade Shaped the World* (New York: Grove, 2008), 333–334.

79. *Gordon's Report*, 156.

80. *Harper's Weekly*, September 30, 1899, 993; Pacyga, *Slaughterhouse*, 16.

81. *Harper's Weekly*, October 14, 1899, 1054.

82. Pacyga, *Slaughterhouse*, 31, 109.

83. Arthur Woodward Booth, "The Preparation of Surgical Catgut," *Therapeutic Gazette* 10, 3rd series, no. 12 (December 15, 1894), 810–819.

84. Pacyga, *Slaughterhouse*, 109.

85. "Nat Landsberger's Tale of Woe: Too Much Rock in Violin Strings," *San Francisco Examiner*, January 6, 1914, 5. Those of us who take care of both sheep and goats occasionally encounter individuals who erroneously believe that goats can digest tin (or aluminum) cans, but this was the first I'd ever heard it said about sheep.

86. *San Francisco Examiner*, January 6, 1914, 5.

87. Ibid.

88. Ibid.

89. Oxford English Dictionary.

90. Alberto Bachmann, *An Encyclopedia of the Violin* (1925; repr., New York: De Capo, 1966), 152.

91. Ibid., 140.

92. Ibid., 141–142.

93. Ibid., 148.

94. "Uses Tons of Catgut: Chicago's Music String Factory the Only One in America," *Sunday Inter-Ocean* (Chicago), September 23, 1900, 36.

95. "Manufacturing Gut Strings," *The Violinist* 21 (April 1917): 146.

96. See David Mannes, *Music Is My Faith: An Autobiography* (New York: Norton, 1938).

97. Frederick H. Martens, *Violin Mastery: Talks with Master Violinists and Teachers* (New York: Stokes, 1919), 157–158. For similar sentiments by other violinists, see ibid., 172, 225–226. See also Elias Dann, "The Second Revolution in the History of the Violin: A Twentieth-Century Phenomenon," *College Music Symposium* 17, no. 2 (Fall 1977): 64–71.

98. Dann, "The Second Revolution in the History of the Violin," 65; Walter Kolneder, *The Amadeus Book of the Violin: Construction, History, and Music*, translated by Reinhard G. Pauly (Portland, Ore.: Amadeus, 1998), 51.

99. Kolneder, *The Amadeus Book of the Violin*, 51–52; Dann, "The Second Revolution in the History of the Violin," 65.

100. A 1915 music store advertisement in the Paducah, Kentucky, *News-Democrat* (August 1, 1915, 11) offered gut strings for ten cents each, while a complete set of four "silvered steel" strings could be had for just five cents.

101. Mannes, *Music Is My Faith*, 269.

Chapter 8. Nothing Is Certain but Death and Taxes

1. "Annual Messages to Congress on the State of the Union (Washington 1790–the Present)," table, American Presidency Project, https://www.presidency.ucsb.edu/node/324107.

2. Grover Cleveland, "Third Annual Message (first term)," December 6, 1887, in *The Campaign Textbook of the Democratic Party of the United States for the Presidential Election of 1888*, National Democratic Committee (New York: Brentanos, 1888), 32.

3. Ibid., 33.

4. Ibid., 32.

5. Ibid., 38.

6. Ibid. Taussig, *The Tariff History of the United States*, 170 ("The woolen schedule

of our tariff is the one which imposes the heaviest and the least defensible burdens on consumers").

7. Cleveland, "Third Annual Message," 36; Joanne Reitano, *The Tariff Question in the Gilded Age: The Great Debate of 1888* (State College: Pennsylvania State University Press, 1994), 94–95.

8. See Henry Hazlitt, *Economics in One Lesson* (New York: Three Rivers, 1979), 75–77.

9. U.S. Constitution, article 2, section 3.

10. W. H. Michael, *Official Congressional Directory*, 3rd ed. (Washington, D.C.: Government Printing Office, 1888), 144, 146.

11. Christopher H. Achen, "Slavery or Sheep? The Antebellum Realignment in Vermont 1840–1860" (paper delivered at the Midwestern Political Science Association Annual Meeting, Chicago, April 2012).

12. Ibid., 6–7.

13. Ibid., 1, 17–18.

14. Republican National Committee, *The Republican Campaign Text-Book for 1884* (New York: Republican National Committee, 1884), 171.

15. Benjamin Harrison, "Letter Accepting the Presidential Nomination," September 3, 1892, American Presidency Project, https://www.presidency.ucsb.edu/documents/letter-accepting-the-presidential-nomination-1. By his use of "people of the West" Harrison pointed to the region of the country known for sheep, not for woolen factories.

16. Reitano, *The Tariff Question in the Gilded Age*, xxii.

17. PTJ, *Retirement Series*, 7:13.

18. Wright, *Wool-growing and the Tariff*, 344–346.

19. Rep. Finances 1887, 43 .

20. "Executive Summary to the 2018 Financial Report of U.S. Government," https://www.fiscal.treasury.gov/files/reports-statements/financial-report/2018/Executivesummary-2018.pdf; "Monthly Treasury Statement," https://fiscaldata.treasury.gov/datasets/monthly-treasury-statement/summary-of-receipts-outlays-and-the-deficit-surplus-of-the-u-s-government; Ulrik Boesen, "Gas Tax Revenue to Decline as Traffic Drops 38 Percent," Tax Foundation, March 31, 2020, https://taxfoundation.org/blog/gas-tax-revenue-decline-as-traffic-drops.

21. Taussig, *The Tariff History of the United States*, 249.

22. Dave Leip's Atlas of U.S. Presidential Elections, map of 1888 election, uselectionatlas.org/.

23. The other states with over a million sheep were California, Oregon, Michigan, Pennsylvania, and Texas; Cleveland carried only Texas. There were also three nonvoting western territories with over a million sheep. See U.S. Census, 1890, *Report on the Statistics of Agriculture*, 236.

24. Ibid., 344.

25. Republican National Committee, *The Republican Campaign Text-Book for 1892* (New York: Republican National Committee, 1892), 125.

26. Ibid.

27. Dave Leip's Atlas of U.S. Presidential Elections, map of 1892 election, uselectionatlas.org/.

28. U.S. Treasury Department, *Wool and Manufactures of Wool* (Washington, D.C.: Government Printing Office, 1888), lxi.

29. See listing and summary in Wright, *Wool-Growing and the Tariff*, 345–346.

30. Cong. Rec. 26, pt. 9 (1894), 418.

31. "Clark, James Beauchamp (Champ)," Biographical Directory of the United States Congress, https://bioguideretro.congress.gov/Home/MemberDetails?memIndex=C000437.

32. Cong. Rec. 26, pt. 1 (1894), 903.

33. Ibid., 902.

34. Ibid., 894–895.

35. Cong. Rec. 26, pt. 10 (1894), 1131.

36. Sharon C. Nantell, "A Cultural Perspective on American Tax Policy," *Chapman Law Review 2* (1999): 33–93, 43–45; Laura F. Edwards, *A Legal History of the Civil War and Reconstruction: A Nation of Rights* (New York: Cambridge University Press, 2015), 30–31; U.S. Stats. 12 (1862), 473–474.

37. U.S. Stats. 28 (1894), 509–570.

38. Ibid., 553.

39. Nantell, "A Cultural Perspective on American Tax Policy," 47.

40. Pollock v. Farmers Loan and Trust Company, 157 U.S. 429 (1895).

41. Akhil Reed Amar, *America's Constitution: A Biography* (New York: Random House, 2005), 405–407; Linda Przybyszewski, *The Republic According to John Marshall Harlan* (Chapel Hill: University of North Carolina Press, 1999), 170–177.

42. Wright, *Wool-growing and the Tariff*, 346.

43. William Howard Taft, inaugural address, March 4, 1909, *American Presidency Project*, https://www.presidency.ucsb.edu/documents/inaugural-address-46. One hundred million dollars in 1909 is equal to nearly $2.8 billion dollars in 2020. *See* Inflation Calculator, Official Data Foundation, https://www.officialdata.org/.

44. William Howard Taft, special message, June 16, 1909, American Presidency Project, *https://www.presidency.ucsb.edu/documents/message-the-congress-concerning-tax-net-income-corporations.*

45. U.S. Constitution, Amendment 16.

46. The vote in favor of the amendment was 77–0 in the Senate and 318–14 in the House; forty-two out of forty-eight states endorsed the amendment, mostly

with huge supermajorities in their respective legislatures. See Nantell, "A Cultural Perspective on American Tax Policy," 49, 90–91.

47. U.S. Stats. 38 (1913), 114–202.

48. Ibid., 142.

49. Ibid., 164–165.

50. Ibid., 166, 167; Nantell, "A Cultural Perspective on American Tax Policy," 50. In 1910 the census recorded a population of 92.4 million people; three years later under the new income tax law there were only 357,598 returns filed. To put this number in perspective, the Census Bureau in 2020 reported that Pulaski County, Arkansas, had nearly 400,000 residents. See David Paris and Cecelia Hilgert, "70th Year of Individual Income and Tax Statistics, 1913–1982," Internal Revenue Service Publication 13–82, https://www.irs.gov/pub/irs-soi/13-82inintrd70yrs.pdf, 1.

51. "Underwood, Oscar Wilder," Biographical Directory of the United States Congress, https://bioguideretro.congress.gov/Home/MemberDetails?memIndex=U000013.

52. John F. Kennedy, *Profiles in Courage* (New York: Harper & Brothers, 1956), 207–208.

53. Cong. Rec. 50, pt. 2 (1913), 1222.

54. *Report of the Commissioner of Agriculture for the Year 1871* (Washington, D.C.: Government Printing Office, 1872), 40.

55. Ibid., where the figure given for Ohio sheep in 1870 is 4,302,904. The 1870 U.S. Census, however, puts the number of Ohio sheep higher, at 4,928,635 (U.S. Census 1870, *Statistics of Agriculture*, 74). Given this discrepancy, I have used the phrasing "fewer than five million" to be in agreement with both sources.

56. *Report of the Commissioner of Agriculture . . . 1871*, 40–41.

57. Cong. Rec. 26, pt. 10 (1894), 830–831.

58. Cong. Rec. 26, pt. 10 (1894), 1355.

59. Cong. Rec. 26, pt. 1 (1894), 577.

60. Cong. Rec. 26, pt. 2 (1894), 1599. See also Reitano, *The Tariff Question in the Gilded Age*, 95.

Chapter 9. A Little Pure Wildness

1. The other species of North American wild sheep is the Dall's sheep, also called the thinhorn sheep, *Ovis dalli*. The two wild American sheep have different ranges. The bighorn inhabits portions of northern Mexico, the continental western United States, and southern British Columbia and Alberta. The thinhorn sheep lives farther north, inhabiting Alaska, the Yukon, and northern British Columbia. See David M. Shackleton, "Ovis canadensis," *Mammalian Species* 230 (1985), 1–9; and R. Terry Bowyer and David M. Leslie Jr., "Ovis dalli," *Mammalian Species* 393 (1992), 1–7.

2. W. J. Foreyt, "Fatal *Pasteurella haemolytica* Pneumonia in Bighorn Sheep after Direct Contact with Clinically Normal Domestic Sheep," *American Journal of Veterinary Research* 50 (March 1989), 341–344. See also Detlef K. Onderka, Shirley A. Rawluk, and William D. Wishart, "Susceptibility of Rocky Mountain Bighorn Sheep and Domestic Sheep to Pneumonia Induced by Bighorn and Domestic Livestock Strains of *Pasteurella haemolytica*," *Canadian Journal of Veterinary Research* 52 (October 1988), 439–444; Detlef K. Onderka and William D. Wishart, "Experimental Contact Transmission of Pasteurella haemolytica from Clinically Normal Domestic Sheep Causing Pneumonia in Rocky Mountain Bighorn Sheep," *Journal of Wildlife Diseases* 24 (October 1988), 663–667.

3. Janet L. George, Daniel J. Martin, Paul M. Lukacs, and Michael W. Miller, "Epidemic Pasteurellosis in a Bighorn Sheep Population Coinciding with the Appearance of a Domestic Sheep," *Journal of Wildlife Diseases* 44 (April 2008): 388–403.

4. Julie Stiver, *Bighorn Sheep Management Plan: Data Analysis Unit RBS-8* ([Denver]: Colorado Division of Wildlife, 2011), 14; J. L. George, R. Kahn, M. W. Miller, and B. Watkins, *Colorado Bighorn Sheep Management Plan, 2009–2019*, Special Report 81 ([Denver]: Colorado Division of Wildlife, 2009), 1.

5. Crosby, *Ecological Imperialism* 195–216; Charles C. Mann, *1491: New Revelations of the Americas before Columbus* (New York: Knopf, 2005), 94–104; Jared Diamond, *Guns, Germs, and Steel* (New York: Norton, 1999), 195–214.

6. *Recommendations for Domestic Sheep and Goat Management in Wild Sheep Habitat* ([Boise]: Western Association of Fish and Wildlife Agencies, 2012), 6.

7. Ibid., 14.

8. Brett Bannor, "Did Lewis and Clark Know That Bighorn Sheep Existed?," *We Proceeded On* 44, no. 4 (November 2018): 15–19. As I discuss in this article, it is common for both popular and scholarly works to state that the first bighorn specimen known to western science was collected by the Lewis and Clark expedition. This is incorrect—the first specimen came from a Canadian surveying expedition that took place nearly four years before Lewis and Clark departed on their journey.

9. Roosevelt, *Hunting Trips of a Ranchman*, 121.

10. See Donald Worster, *A Passion for Nature: The Life of John Muir* (New York: Oxford University Press, 2008). See also Andrea Wulf, *The Invention of Nature: Alexander von Humboldt's New World* (New York: Knopf, 2015), 315–334; Aaron Sachs, *The Humboldt Current: Nineteenth Century Exploration and the Roots of American Environmentalism* (New York: Penguin, 2006), 305–332.

11. John Muir, January–May 1869 journal, John Muir Papers, Holt Atherton Special Collections—Digital Archives, University of the Pacific, image 5 and image 20, https://scholarlycommons.pacific.edu/jmj3.

12. Worster, *A Passion for Nature*, 158–159.

13. John Muir, "The Mountain Lakes of California," *Scribner's Monthly* 17, no. 3 (January 1879): 411–420, 416. The text in *Scribner's* actually reads "hooped locusts" rather than "hooved locusts," but Donald Worster persuasively remarks that this is surely a misprint. See Worster, *A Passion for Nature*, 474n18. "Money changers in the temple" is a reference to John 2:13–16; sheep were among the things the money changers were selling in the temple.

14. John Muir, "My First Summer in the Sierra," *Atlantic Monthly* 107, no. 4 (April 1911): 521–528, 523. See also Worster, *A Passion for Nature*, 160–162, 312.

15. See Worster, *A Passion for Nature*.

16. John Muir, "The Wild Sheep of California," *Overland Monthly*, April 1874, 358–363, 359.

17. John Muir, "Wild Wool," *Overland Monthly and Out West Magazine*, April 1875, 361.

18. Ibid, 363.

19. Ibid., 365.

20. Ibid, 366.

21. Ibid.

22. U.S. Stats. 13 (1864),325. See also Adam Wesley Dean, *An Agrarian Republic: Farming, Antislavery Politics, and Nature Parks in the Civil War Era* (Chapel Hill: University of North Carolina Press, 2015), 108–134.

23. Ibid., 130.

24. U.S. Stats. 26 (1890),650–651.

25. Kimberly K. Smith, *The Conservation Constitution: The Conservation Movement and Constitutional Change, 1870–1930* (Lawrence: University Press of Kansas, 2019), 2–3; Dean, *An Agrarian Republic*, 7–8, 120–121.

26. Alexis de Tocqueville, *Democracy in America*, translated by Arthur Goldhammer (New York: Library of America, 2004), 310.

27. U.S. Stats 26 (1891), 1095–1103; Smith, *The Conservation Constitution*, 99–101.

28. Benjamin Harrison, "Proclamation 348—Setting Apart as a Public Reservation Certain Lands in the State of California," February 14, 1893, American Presidency Project, https://www.presidency.ucsb.edu/documents/proclamation-348-setting-apart-public-reservation-certain-lands-the-state-california. See also Brinkley, *The Wilderness Warrior*, 238.

29. "John Muir Wilderness," SierraWild.gov, https://www.sierrawild.gov/wilderness/john-muir.

30. U.S. Stats. 30 (1897),35.

31. *United States v. Blasingame*, 116 F. 654 (S.D. Cal. 1900).

32. Smith, *The Conservation Constitution*, 123.

33. House of Representatives, 58th Cong., 2nd sess., *Grazing on Forest Reserves* (Report No. 1148, 1904), 2.

34. U.S. Stats 33 (1905), 628.

35. *United States v. Grimaud,* 220 U.S. 506, 509 (1911); Smith, *The Conservation Constitution,* 121–127.

36. U.S. Stats 34 (1907), 1246; Smith, *The Conservation Constitution,* 123.

37. *United States v. Grimaud,* 220 U.S. 515.

38. Ibid., 510.

39. Ibid., 517.

40. Ibid., 516. See also Smith, *The Conservation Constitution,* 121–127; Char Miller, *How Counting Sheep Saved the U.S. Forest Service,* Forest History Society, May 3, 2011, https://foresthistory.org/how-counting-sheep-saved-the-u-s-forest-service.

41. Thomas W. Merrill and Kathryn T. Watts, "Agency Rules with the Force of Law: The Original Convention," *Harvard Law Review* 116, no. 2 (December 2002): 501–502.

42. Smith, *The Conservation Constitution,* 126.

Chapter 10. Swimsuits, Soldiers, Polyester

1. Jean Kincaid, "Nothing Fancy, You Know, but Bathing Suits for the Every-day Woman," *Boston Weekly Globe,* August 2, 1890, 1.

2. Ibid.

3. Advertisement in *Vogue,* April 1, 1949, 59.

4. Donald MacGregor, "Will Wool Win the War?," *Los Angeles Times,* February 8, 1942, 102–103.

5. Ibid, 103.

6. Ibid.

7. John W. Klein, *Wool during World War II,* War Records Monograph 7, (U.S. Department of Agriculture, Bureau of Agricultural Economics, 1948), 13. Recall that during the Civil War, Union soldiers had an annual consumption of about sixty pounds per soldier. Wright, *Wool-Growing and the Tariff,* 173.

8. Klein, *Wool during World War II,* 8–9. On "up to a billion pounds" estimate, see also *Hearings before the Special Committee to Investigate the Production, Transportation, and Marketing of Wool,* U.S. Senate, 79th Cong., 1st session, pt. 6 (1945), 1109, hereafter *Hearings to Investigate Wool.*

9. Klein, *Wool during World War II,* 9.

10. Ibid., 12.

11. Connor's proposal is in *Hearings to Investigate Wool,* 1021–1029.

12. Ibid., 1021.

13. Ibid., 1023–1024.

14. Klein, *Wool during World War II*, 7.

15. Ibid., 22.

16. Ibid., 17.

17. Ibid., 6.

18. Ibid 20–21. For texts of these agreements, see *Hearings to Investigate Wool*, 1112–1122.

19. Klein, *Wool during World War II*, 18.

20. Ibid., 8, 61.

21. Leslie Carpenter, "Probers Blame Munitions Board for Wool Shortage," *Fort Worth Star Telegram*, December 10, 1950, 13.

22. *Third Report of the Preparedness Subcommittee of the Committee on Armed Services, United States Senate: Agricultural Products and the Mobilization Program*, (Washington: Government Printing Office, 1950).

23. Ibid., 8.

24. Ibid., 8–9. Furthermore, efforts to buy foreign wool hit a snag when the United States approached the Australian government, seeking a discount in exchange for buying all of that nation's available wool. Australia refused, insisting that America would have to pay the market price. See Evan McHugh, *The Shearers: The Story of Australia, Told from the Woolsheds* (Australia: Penguin Group, 2015), 199–200.

25. *Third Report of the Preparedness Subcommittee*, 22.

26. Ibid., 21–22.

27. Ibid., 2.

28. Jerry Greene, "1 1/2 Billion for Stockpile, Not Enough Wool for GIs," *New York Daily News*, December 16, 1950, 139.

29. *Third Report of the Preparedness Subcommittee*, 22.

30. *Third Report of the Preparedness Subcommittee*, 10.

31. David A Hounshell and John Kenly Smith Jr., *Science and Corporate Strategy: Du Pont R&D, 1902–1980* (Cambridge: Cambridge University Press, 1988), 175, 325, 391, 392.

32. Ibid., 392.

33. The amount of $70 million dollars in 1930 is equal to over a billion dollars in 2020. Inflation Calculator, Official Data Foundation, https://www.officialdata.org/.

34. Hounshell and Smith, *Science and Corporate Strategy*, 257–273; J. E. McIntyre, "Historical Background," in *Synthetic Fibres: Nylon, Polyester, Acrylic, Polyolefin* (Cambridge, UK: Woodhead, 2005), 8.

35. Hounshell and Smith, *Science and Corporate Strategy*, 264.

36. Ibid., 407–420.

37. John T. Molloy, *John T. Molloy's New Dress for Success* (New York: Warner Books, 1988), 70.

38. Keithly G. Jones, "Trends in the US Sheep Industry," Agricultural Information Bulletin no. 787 (2004), iii.

39. Luke Runyon, "The Long, Slow Decline of the U.S. Sheep Industry," Nebraska Public Media, October 15, 2013, https://nebraskapublicmedia.org/en/news/news-articles/the-long-slow-decline-of-the-us-sheep-industry.

40. U.S. Tariff Commission, *The Wool-Growing Industry*, 588.

41. Ibid., 554, 567. See also Mark A. Smith, *The Tariff on Wool* (New York: Macmillan, 1926),117.

42. Melane Salque, Peter I. Bogucki, Joanna Pyzel, Iwona Sobkowiak-Tabaka, Ryzard Grygiel, Marzena Azmyt, and Richard P. Evershed, "Earliest Evidence for Cheese Making in the Sixth Millennium B.C. in Northern Europe," *Nature* 493 (2013): 522–525.

43. Of course, sheep, like cattle, can also be a source of dairy products, as can goats and some other ruminants. In the discussion here, however, I'm equating dairy specifically with bovines, since they provide the bulk of production in the United States.

44. U.S. Census 1920, Abstract, 16.

45. Just six years earlier, the conservationist William T. Hornaday had published an influential treatise discussing with great alarm the destruction of America's wild animals. See *Our Vanishing Wildlife: Its Extermination and Preservation* (New York: Scribner, 1913).

Epilogue. "The Sheep Follow Him, for They Know His Voice"

1. See *The Jefferson Bible: The Life and Morals of Jesus of Nazareth* (Mineola, N.Y.: Dover, 2006). See also Steven Waldman, *Founding Faith: How Our Founding Fathers Forged a Radical New Approach to Religious Liberty* (New York: Random House, 2008), 72–85; and Edwin S. Gaustad, *Sworn on the Altar of God: A Religious Biography of Thomas Jefferson* (Grand Rapids, Mo.: Eerdmans, 1996), 123–131. For a refutation of the claim that the Jefferson Bible expunged all miracles, see Daniel L. Dreisbach, *Reading the Bible with the Founding Fathers* (New York: Oxford University Press, 2017), 63–64.

Index

Locators in italics indicate a figure.

Achen, Christopher H., 152
Adams, Abigail, 14, 198n32
Adams, John, 14, 198n32
Adams, John Quincy, 14, 48
African sheep, 57, 209–10n4
Ambrose, Stephen, 104
American West: myths about, 113–14; presence of sheep in, 111–12, 189–90; settlement and, 100–106; shearer strikes in, 126–27; sheep husbandry and, 230n15. *See also Big Valley, The* (show); conservation; Frontier Thesis; Muir, John; sheepmen-cattlemen dichotomy; *Virginian, The* (Wister); *individual states*
Angora goats, *122,* 130, 132
antislavery movement and wool, 152
Arator (Taylor), 30–33
Arlington long-wooled sheep, 26
Armour & Company, 147–48
Articles of Association, 12, 13, 27
Azores Islands, 9, 10

Bachmann, Alberto, 146
Ball, Charles, 58, 65
Barry, Samuel F., 37, 39
Beatty, Adam, 37–38
Benjamin Franklin (ship), 24
Benton, C., 37, 39

Benton-Barry sheep census, 37
Berkshire Agricultural Society, 29–30
bighorn sheep, 163–66, 169–70, 173, 221n68, 233n1, 234n8
Big Valley, The (show), 109–11
Blackstone, William, 77
blankets, 15, 20. *See also* woolens
Blenheim Palace, 133–34
Bond, Scott, 74
Bowers, William, 156
Brinkley, Douglas, 112
Britain. *See* wool industry, British
Brown, John, 65–67
Budding, Edwin Beard, 133
Buenos Aires wool, 41–42, 46, 49
Burden, Charles, 86–87, 89–91
Burden v. Hornsby, 85–91, 216n40

calling of sheep, 191–92
candles, 145
carcass, uses of. *See* catgut; meat; soap
cars and mohair, 130, 132
catgut, 145–49
cattle, 32, 102–4, 188
cattle culture, 100–105, 107–11, 112–14. *See also* sheepmen-cattlemen dichotomy
cattlemen-sheepmen feuds, 84–85, 111
Cawthon, Cicely, 34, 50–51
Central Park (New York City), 134–36
Charch, William Hale, 184, 186–87

239

Cherokee Nation, 3
Chicago, IL, 142–43, 146, 187
Choate, Julian Ernest, Jr., 113–14
civilization and sheep, 5–6, 31, 68–69, 196n13
Clark, James Beauchamp, 157, 159
Cleaver, Samuel, 140–41
Cleveland, Grover, 150–53, 154–55
clothing. *See* woolens
coarse wool, 41–44, 47–49, 206n38
Colfax courthouse massacre, 70–71
Colonial North America, 9–14
colonization, 9–10, 16
Columbian Exchange, 8–9, 164
Columbian Exposition (World's Fair), 106
Columbus, Christopher, 8–9, 195n7
Commentaries on the Laws of England (Blackstone), 77
Commodity Credit Corporation (CCC), 182–83
Connecticut, 37
Connor, Louis G., 179
conservation, 166, 168, 170–73
cotton, 31, 33, 40–41, 49, 58–59
Coxe, Tench, 36–37
coyotes, 168
Crèvecoeur, J. Hector St. John de, 7
criticism of sheep culture, 31–32, 40–41. *See also* sheepmen-cattlemen dichotomy
cruelty to animals, 223–24n7

dairy, 187–89, 203n70, 237n43. *See also* milking of sheep
Daisy (sheep), 189
Dall's sheep, 233n1
Darwin, Charles, 18–19
Davis, David Brion, 59
Decline and Fall of the Roman Empire, The (Gibbons), 26
destruction of sheep, 14–15, 161
disease, 163–64
dogs: Central Park sheep and, 135–36; destruction of, 85–86, 91, 93; enslaved ownership of, 80; "fice," 214n6; legislation on, 75–77, 80–82, 85–89, 90–93; owner accountability for, 79; sheepdogs, 73, 78; threat of, 63, 68–69, 75–80, 84–86, 213n58; use of, 73; "value" of, 78. *See also* dog taxes
dog taxes, 81–85, 91–92
domestication of sheep, 3, 5–6, 9, 31, 35, 43. *See also* civilization and sheep; sheep husbandry; sheep improvement
domestic-wild sheep: comparisons (Muir), 169; management, 163–64, 166, 170
Don Pedro (sheep), 24, *25*
drives of livestock, 105
Du Bois, W.E.B, 73–74
du Pont, E. I., 24, 51–52, 186
DuPont Company, 184–86

economic policy and wool: American, 20–23; British, 21; Merinos and, 27; *premiums* and *bounties* in, 201n25; sheep as "species of capital," 41; Spanish, 24–25. *See also* taxation and revenue collection; wool tariffs
"effective separation," 166, 170
embargoes, 17–18, 28
enslaved people: dogs and, 80, 92–93; as shepherds, 57–59; South Carolina and, 61; wool consumption of, 38, 41–44, 205n23
Essay on Sheep (Livingston), 1–2, 5, 15–16, 26, 195n1, 196n13
"Eulogy for the Dog" (Vest), 87–90

fairs, 29–30
Federal Writers' Project (FWP), 34–35
feminization of sheep culture, 100, 113–17. *See also* "Mary's Lamb" (Hale)
fences, 72–73
Finley, Ruth, 97, 99, 219n25
firearms, 69–71
First Continental Congress, 12, 13, 27, 142
Ford, Henry, 96–100, 128, 130, 219n25
"Ford Cycle of Production" (World's Fair exhibit), 130–32
Ford Motor Company, 128, *131*, 132, 225n30

Ford sheep film, 128–30, 225n30
forest reserves, 171–72
Foreyt, William, 163–64
Foreyt sheep study, 163–64
Fourteenth Amendment, 70–71
Franklin, Benjamin, 10–12, 14, 60
Franklin Park (Boston), 132–33, 134, 136–37
Frantz, Joe B., 113–14
Frontier Thesis, 106–7, 111–12, 114
fur trade, 2, 13

gender bias and sheep culture, 113–14. *See also* feminization of sheep culture; masculinity
Georgia, 38, 92. *See also Miller v. State*
Gibbon, Edward, 26–27, 202n44
globalism and wool, 49
goats, *122*, 130, 132, 229n85
Gordon, Clarence W., 102, 104–6
grazing, 40–41, 104, 166, 168–69, 171–73. *See also* lawn maintenance; rangeland
grazing permits, 171–73
Great Plains, 101–2, 104, 108
Grey's Raid, 14
Gulf Coast sheep, 9
guns. *See* firearms
Guy, Francis, 3, *4*

Hahn, Steve, 35
Hale, Richard, 97–99
Hale, Sarah Josepha, 95–99
Hamilton, Alexander, 13, 20–22, 23, 28, 201n25
hardiness of sheep, 104–5
Harding, Warren, 139
Harrison, Benjamin, 153, 154–55, 171
Hayes, John Lord, 67–68
Hayes report, 68–69
Herbemont, Nicholas, 60–62, 64–65, 69
homespun *versus* factory-made, 18, 35, 44–45, 51. *See also* woolens
Hornsby, Leonidas, 85–87, 89–91
horses, 113–14, 212n54
House of Commons, 10–11

Houx, Frank L., 126
Howe, Daniel Walker, 52
hunting, 113. *See also* Roosevelt, Theodore

importation of wool in US: Australia and, 178–79, 236n24; Britain and, 15–18, 198n26; Buenos Aires wool and, 41–42, 46, 49; coarse wool and, 207n52; globalism and, 49; Merinos and, 24–25, 201n35; pre-Civil War, 43; Preparedness Subcommittee report on, 182; quantities of, 42–43; reliance on, 12, 13; Smyrna wool and, 41–42, 46, 49. *See also* military demand for wool; wool tariffs
income tax, 157–62, 232n50
industrialization. *See* lawn maintenance; meat; refrigeration; shearing machines
insurrections, 60–61

Jacob's sheep (biblical story), 19, 200n9
James, Mary, 34, 50
Jarvis, William, 25
Jefferson, Thomas: bible of, 191–92; on Blenheim Palace, 133; defense of sheep, 31–32; on dogs, 79–80, 82–83; Merino sheep and, 22–23, 25–28, 153, 202n48; sheep husbandry and, 35, 51, 57–58; sheep improvement, 25–28, 30; wool quality and, 81, 204n6
Johnson, Lyndon B., 182–84
John the Shepherd. *See* Brown, John
Jungle, The (Sinclair), 111

King, J. Crawford, 72–73

Lamar, Joseph, 172
lamb (meat), 108, 140–42, 187
Landsberger, Nathan J., 145–46
lawn maintenance, 133–34, 135, 137
lawn mowers, 133–34, 136, 226n36, 226n40
Lee, Daniel, 63–65, 189
Letters from an American Farmer (Crèvecoeur), 7–8, 196–97n4

Index 241

Libby Company, 139–40, 142, 144
Livingston, Robert R.: *Essay on Sheep*, 1–2, 5–6, 15–16, 195n1, 196n13; Merino sheep and, 24; patriotism of, 26, 202n42; sheep improvement and, 201n39
looms, 5, 35, 50

Madison, James, 1, 27, 28, 202n48
Mannes, David, 148–49
manure, 30–31, 203n70
Martha's Vineyard, 7, 8, 10, 14
"Mary's Lamb" (Hale), 94–95, 97–100, 218n3
masculinity, 100, 112–14. *See also* feminization of sheep culture; gender bias and sheep culture
meat, 108, 139–42, 144, 187, 228n70
meatpacking industry, 139–40, 142–43
Merino sheep: disposing of, 202n48; dog taxes and, 82; importation of, 24–26, 207n58; meat of, 142; microscopic image of wool of, 184–85; patriotism and, 202n46; Pittsfield fair and, 28–30; sales of, 26–27; sheep improvement and, 23, 28, 201n39; Spanish restrictions on, 24–25, 201n35; US wool industry and, 51–52; wool uses and, 81
Mesta (Spain), 24
"Middle West, The" (Turner), 108
military demand for wool: civilization and, 5–6; Civil War and, 50, 53–55, 135–36, 158, 161, 236n7; Cold War and, 182; deficit concerns and, 156, 182–84; executive power and, 23; Korean War and, 182–83; masculinity and, 117; Revolution and, 14–15, 23; War of 1812 and, 28; World War I and, 115, 117, 125–26, 224–25n20; War World II and, 177–81. *See also* war efforts
milking of sheep, 203n70
Miller, Daegan, 67
Miller v. State, 75–77, 78, 85, 93
Missouri. *See Burden v. Hornsby*
mohair, 130, 132
Montana, 104–5, 221n66

Muir, John, 166–70
Munitions Board, 182–83
music, 149. *See also* violin strings
mutton, 108, 140–43. *See also* meat; meatpacking industry

Nantucket, 7–8, 10
National Association of Wool Manufacturers, 67
"Negro cloths," 42–44, 46, 48, 49, 52
"Negro Farmer," (Du Bois), 73
New York. *See* Central Park (New York City)
Nicholson, Jospeh, 17
North Dakota, 100–101, 104
Notes on the State of Virginia (Jefferson), 35, 51–52, 64
nylon (Fiber 66), 185

Old Drum (dog), 86–87, 88, 91
Olmstead, Frederick Law, 80, 132–37
Origin of Species, The (Darwin), 18–19
orphaned lambs, 94–95
ownership of sheep, 56–57. *See also* sheep farmers

Panic of 1819, 46
Parable of the Good Shepherd, 191–92
Parliament, 10–11, 21
pasturage, 8
patriotism, 26–27, 67, 138–39, 202n46
Pennsylvania, 78–79, 82
Peters, Richard, 26, 78–79, 130
pilfering of animals, 68, 72, 77–78
Pittsfield Fair, 29
plantation economy and wool, 34–39, 42–47, 49–53, 57, 65, 208n78
Pleasant Valley War, 111
Poems for Our Children (Hale), 96–98
politics: agriculture and, 30–31; conservation and, 170–73; sheep populations and, 152–55; wool tariffs and, 154–57, 160
Pollock v. Farmers Loan and Trust Company, 159

polyester (Fiber V), 186
popular culture representation of sheep. *See Big Valley, The* (show); feminization of sheep culture; *Jungle, The* (Sinclair); *Virginian, The* (Wister)
populations of sheep: American Revolution and, 15; Civil War and, 54; destruction of surplus sheep and, 161; meat consumption and, 141; in Northeast, 36, 115; politics and, 152–55; reporting of, 36–37, 39, 40, 102–6, 187, 204–5n13, 204n10; in South, 35, 38, 62; in twenty-first-century U.S., 189–90; US decline of, 187; in West, 102–5, 109–10, 221n66, 222n72, 231n23
Powell, Arthur Gray, 76–77, 78, 90, 93
predation concerns, 84–85, 168. *See also* coyotes; dogs; wolves
Preparedness Subcommittee report, 182–83
protection of sheep. *See* dogs; firearms
public virtue, 12–13, 27. *See also* patriotism
Pugh, Nicey, 34, 50

race and sheep husbandry, 60–64, 68–69. *See also* plantation economy and wool; shepherds
Randall, Henry S., 38, 44, 59, 205n23, 206–7n44
Randolph, Mary, 142
rangeland, 72–73, 104, 107. *See also* grazing
rats, 9
ready-made clothing, 18, 35, 44–45, 51. *See also* woolens
Reconstruction, 68, 70–73, 92
Redstone Schoolhouse, 97, 98
refrigeration, 143, 188
Reibelt, J. Phillipe, 22–23
Reitano, Joanne, 153
Reynold, David S., 66
Rickey, Don, Jr., 101
Riley, Glenda, 114
Rondalson, James, 81–82
Roosevelt, Theodore, 101–2, 104, 106, 108, 166–68

Roosevelt, Theodore (works): *Hunting Trips of a Ranchman,* 100–101; "The Northwest in the Nation," 111–12; *The Wilderness Hunter,* 113; *The Winning of the West,* 106
rural-urban perception of sheep, 114–15

Saxony sheep, 65, 142
scarcity of sheep, 141
Seabury, Samuel ("A. W. Farmer"), 13
Second Continental Congress, 60
selective breeding. *See* sheep improvement
sharecropping, 69
shearer labor strikes, 126–27
shearing and shearers: animal cruelty and, 223–24n7; evolution of, 123–25; "Ford Cycle" exhibit and, 130; Ford film and, 128–30; labor needs and, 59; manual *versus* mechanical, 224n8, 224n10; modern processes for, 118–20; records of, 224n10; Sheep Shearers Union of North America, 225n23. *See also* shearing machines
shearing machines: cars and, *122*, 127–28; Ford film and, 128–30; *versus* hand shearing, 123–25; horse-powered, 121; modern electric, 118, *120*, 121; "Stewart Little Wonder," *121*; Wolseley, 123
"Sheep Club" poster, 115–17
sheepdogs, 73, 78. *See also* dogs
sheep farmers: African American, 69, 214n74; attentiveness of, 18–19; dairy and, 189; labor costs and, 127; Muir and Roosevelt on, 166, 168; "prime" sheep and, 41; shearing wages and, 126; tariffs and, 151–55, 158, 162, 180; war efforts and, 54; wool quality and, 52–53. *See also* dogs; sheep husbandry; shepherds
sheep husbandry: African Americans and, 66, 68–69, 73–74; catgut and, 145–46; civilizing influence of, 68–69; compared with dairy farms, 188–89; critiques of, 31–32; Early American, 9–10, 12–13; flock establishment and, 9, 10; guns and, 69, 71; as hobby, 1; islands and, 9, 10; John Brown

sheep husbandry *(continued)*
and, 65–66; John Muir and, 168–69; labor needs for, 56–59, 210n11; public virtue and, 12–13; range land and, 72; rearing and, 94; reports on, 84–85, 166, 212n52; "Sheep Husbandry by the Colored Population" (Hayes), 68; *Sheep Husbandry in the South* (Randall), 38, 44, 59, 205n23, 206–7n44; in South, 63, 68, 69; war efforts and, 179–80; wool quality and, 43–44. *See also* grazing; plantation economy and wool; sheep improvement; shepherds

sheep improvement: Articles of Association and, 12; Darwin on, 18–19; dog taxes and, 82–83; domestication and wool quality in, 43–44; Jacob's sheep and, 200n9; meat and, 142; Merino sheep and, 24, 27; Muir on, 169–70; patriotism and, 26; policy on, 20–21, 22; principles of heredity and, 19; private, 19–20; Spanish Mesta and, 24; Washington flock and, 19–20; wool imports and, 199n42

Sheep Meadow (Central Park), 135–36

sheepmen. *See* feminization of sheep culture; sheep farmers; sheep husbandry; sheepmen-cattlemen dichotomy; shepherds

sheepmen-cattlemen dichotomy: in early US, 32; feuds and, 84–85, 111; livestock populations and, 100–106; masculinity and, 112–14; in popular culture, 106–12

shepherdesses, 114–17

shepherds: African American, 58–59, 63, 68–69, 72–74, 92–93, 214n75; calling of sheep by, 191–92; female, 114–17; Roosevelt opinion of, 101; transportation of, 114; white, 61–62. *See also* Brown, John; Muir, John

Sierra Nevada mountains, 168–69, 171–72

"Significance of the Frontier in American History" (Turner), 106–7, 112, 114

Sinclair, Upton, 111

Sixteenth Amendment, 159–61

slaughterhouses, 143, 228n77

slavery: property in people and, 59–60; sheep husbandry and, 61–67; sheep *versus* cotton and, 56–69; wool consumption and, 38–39, 42–45, 47, 50–51, 53, 65

slavocracy, 45, 49, 52

Smith, Adam, 20–21, 200n18

Smyrna wool, 41–42, 46, 49

soap, 144–45

South: dependence on Northern wool, 45, 47; gun laws in, 70–71; objections to sheep in, 40–41; race-based laws in, 73; rangeland in, 72–73; sheep husbandry in, 32–33, 61–64, 68–69; sheep populations in, 35; wool consumption in, 42–44, 53. *See also* individual states

South Africa, 178–79

South Carolina, 49, 60–62

Spain. *See* Merino sheep

Stamp Act, 10

Story of Mary and Her Little Lamb (Ford), 98, 99

strikes by shearers, 126–27

Strong, James, 41

sweaters, 175

Swift and Company, 143–44

swimwear, 174–77. *See also* woolens

synthetic fibers, 184–87

Taft, William Howard, 159

Tariff Bill of 1828 (Tariff of Abominations), 39–41, 44, 47–49, 52–53, 205n27, 206n30. *See also* wool tariffs

Tariff Commission, 125–27, 140–41, 178–79, 187–88

tariff laws, 39–41, 44, 46–49, 52–53, 205n27, 206n30. *See also* wool tariffs

Tariff of Abominations. *See* Tariff Bill of 1828 (Tariff of Abominations)

Taussig, Frank, 154

taxation and revenue collection, 150–53, 157–60, 161–62. *See also* dog taxes; income tax; Tariff Bill of 1828 (Tariff of

Abominations); tariff laws; wool tariffs
Taylor, John (of Caroline), 30–33, 40, 203nn68–70
textiles, 50, 184–85. *See also* woolens
theft of animals. *See* pilfering of animals
Thompson, Wiley, 53
threats to livestock, 9
Tocqueville, Alexis de, 171
tods (measure of wool), 21
transatlantic slave trade, 57, 63, 211n35
Treasury Department, 42, 46, 48, 150, 153, 206–7n44
Treaty of Basle, 24
Tunis sheep, 26
turf. *See* lawn maintenance
Turner, Daniel, 44–45, 47
Turner, Frederick Jackson, 106–8, 111–12, 114
turnips, 32

Underwood, Oscar, 160
Underwood-Simmons Tariff Act, 160–61
Union Stockyards, 143
United States Department of Agriculture (USDA), 84
United States v. Blasingame, 171–72
United States v. Cruikshank, 71
United States v. Grimaud, 172–73
urban-rural perception of sheep, 114–15

Vermont, 152
Vesey, Denmark, 60–61
Vest, George Graham, 87–90
violin strings, 145–49, 229–30n100
Virginia, 27, 32–33, 79, 82–84
Virginia House-wife, The (Randolph), 142
Virginian, The (Wister), 108–9, 112, 221–22n70, 221n68, 223n89

war efforts: sheep husbandry and, 135–36, 138, 179–80; youth sheep farmers and, 179–80. *See also* military demand for wool
Washington, George: Cherokees and, 3; on dogs, 79–80; on enslaved shepherds, 210n12; flocks of, 19–20, 58; wool shortages and, 14–15, 183–84; wool yields and, 200nn11–12
Watson, Elkanah, 28–29
Wayman, Alexander W., 56, 58, 209n1
Wealth of Nations, The (Smith), 20–21, 200n18
West. *See* American West
Western Association of Fish and Wildlife Agencies, 164–65
westward settlement, 105–8
White House sheep, 137–39, 227n57
Whitting, Anthony, 19–20
wildlife management, 164–66
"Wild Sheep of California" (Muir), 169
Wild Sheep Working Group, 166
"Wild Wool" (Muir), 169–70
Wilentz, Sean, 46
Wilson, Woodrow, 137–38, 159
Wilson-Gorman Tariff Act, 158
wine, 60, 62
Winter Scene in Brooklyn (Guy), 3–5
Wisconsin, 107
Wister, Owen, 108–12, 115, 221–22n70, 221n63, 223n89
Wolseley, Frederick York, 123, 127–28
Wolsely Company, 223n2
wolves, 9–10, 73, 79, 83, 85, 197n14
Wood, Gordon, 13
wool: advantages of, 2–3, 5, 188; British flocks and, 15–16; Central Park sheep and, 134; dairy and, 188–89; decline of, 177; export regions, 41–42; Ford cars and, 130; foreign dependency on, 12, 13; global sources of, 177–78; goats and, 130, 132; government control of, 126; *versus* meat, 141–42; necessity of, 188–89; oceanic voyages and, 3; plantation economy and, 34–35; Renaissance Europe and, 2–3; representation in art, 3–5; structure of, 184–85; surplus liquidation, 181; White House sheep and, 138; wild, 169. *See also* importation of wool in US; Merino sheep; military demand for wool;

wool *(continued)*
	plantation economy and wool; woolens; wool industry, American; wool quality
wool auction (White House sheep), 138–39, 227n57
wool consumption, 38, 41–44, 205n23, 224–25n20, 236n7. *See also* military demand for wool
wool deficit concerns, 156, 182–84
woolens: blankets, 15, 20; Civil War uniforms and, 53–55; cowboy clothing, 101–2; factory-produced *versus* homespun, 18, 35, 44–45, 51; jean cloth and, 55; matchcoats and, 3; sweaters, 177; swimwear, 174–76. *See also* textiles
woolgrowers. *See* sheep farmers
wool industry, American: American sheep and, 40; British influence on, 45; Civil War era, 50–51, 54; Colonial, 10–12; critique of, 31–32; decline of, 187–88; in Early Republic, 18, 28; economic policy and, 20–22, 52; homespun *versus* factory-made in, 51; imports and, 42–43; Merino sheep and, 25–26; mills and factories in, 51–52; Panic of 1819 and, 46; politics and, 155–56; public virtue and, 12–13; reports on, 20–22; Revolutionary era, 15; South and, 33, 44–48, 52; U.S. Constitution and, 22–23; woolen mills and, 51; wool quality and, 189, 204n6; World War II and, 181. *See also* plantation economy and wool; wool tariffs
wool industry, British, 15–16, 21
wool industry, Italian, 2–3
wool industry, Spanish, 24–25
wool quality, 41–44; in Americas, 11; George Washington and, 200n12; John Brown and, 65; Thomas Jefferson and, 81; wool type and, 41–42
wool soap, 144–45
wool stockpiles, 181–83
wool tariffs: Democratic party and, 154–55, 160; duty payments and, 208n76; income tax and, 157–62; live animals and, 207n58; politics and, 150–57, 160; protection *versus* revenue in, 52–53; for revenue collection, 153; sheep destruction and, 161–62; Sixteenth Amendment and, 160–61; Tariff Bill of 1828, 39–42, 44–49, 52–53, 205n27, 206n30
Wright, Chester Whitney, 51, 54, 154
Wright, Silas, 36, 39, 43, 44, 204n10
Wyoming, 108–9, 125–27, 221n66

Yosemite National Park, 168, 170
youth sheep farmers during World War II, 179–80